THE RARE METALS WAR

Guillaume Pitron, who was born in 1980, is a French award-winning journalist and documentary-maker for France's leading television channels. His work focuses on commodities and on the economic, political, and environmental issues associated with their use. *The Rare Metals War* is his first book, and has been translated into eight languages. Guillaume Pitron holds a master's degree in international law from the University of Georgetown (Washington, DC), and is a TEDx speaker. More information at www.guillaumepitron.com.

Bianca Jacobsohn is a South African/French translator and conference interpreter who specialises in energy, finance, strategic metals, and diplomacy. More information at www.biancajacobsohn.com.

the dark side of clean energy and digital technologies

GUILLAUME PITRON

Translated by Bianca Jacobsohn

Scribe Publications
2 John St, Clerkenwell, London, WC1N 2ES, United Kingdom
18–20 Edward St, Brunswick, Victoria 3056, Australia
3754 Pleasant Ave, Suite 100, Minneapolis, Minnesota 55409, USA

First published in French in 2018 by Les liens qui Libèrent as *La guerre des métaux rares*

First published by Scribe 2020
Reprinted 2021 (twice)

Typeset in Fairfield LT Std by the publishers

Printed and bound in United States of America

Scribe is committed to the sustainable use of natural resources and the use of paper products made responsibly from those resources.

978 1 950354 31 3 (US edition)
978 1 912854 26 4 (UK edition)
978 1 925849 32 5 (Australian edition)
978 1 925938 60 9 (ebook)

Catalogue records for this book are available from the National Library of Australia and the British Library.

scribepublications.com
scribepublications.co.uk
scribepublications.com.au

To my father,
to my mother

CONTENTS

Foreword

by Hubert Védrine
French minister of foreign affairs under President Jacques Chirac, and secretary general and diplomatic adviser to French president François Mitterand

IN AN INCISIVE AND TROUBLING ACCOUNT, GUILLAUME PITRON SOUNDS the alarm on a serious geopolitical problem: the world's growing reliance on rare metals for its digital development in information and communication technologies. This includes the manufacture of devices such as mobile telephones, not to mention the much-lauded electric and/or hybrid car, which requires twice as many rare metals as the humble internal-combustion engine vehicle.

There is nothing untoward about these thirty or so rare metals bearing perfectly civilised Latin names like 'promethium'. They are found in minute proportions in more abundant metals, making their extraction and refinement expensive and difficult. The first problem is that most of these resources are in the hands of China — an advantage it is naturally tempted to exploit. Other countries with such underground resources have for various reasons abandoned their mining operations, which largely gives China a global monopoly and Beijing the title of the 'New Rare Metals Master'.

Pitron illustrates the perils of this dependence with numerous case studies — ranging from super magnets to long-range missiles — where the West has acted inconsistently or entirely without foresight. The solution seems obvious: reopen rare metal production in the United States, Brazil, Russia, South Africa, Thailand, Turkey, and even in the 'dormant mining giant' of France.

Enter the next predicament: mining these rare minerals is anything but clean! Says Pitron, 'Green energies and resources harbour a dark secret.' And he's quite right: extracting and refining rare metals is highly polluting, and recycling them has proved a disappointment. We are therefore faced with the paradox that the latest and greatest technology (and supposedly the greenest to halt the ecological countdown) relies mostly on 'dirty' metals. Thus, information and communication technologies actually produce 50 per cent more greenhouse gases than air transport! It's an especially vicious circle.

How do we overcome the contradiction?

We need to revive the mining of rare earths and of mineral resources internationally (potentially reviving tensions between governments and mining companies), but in an environmentally sound way, using the latest financing, innovation, and other economic and technological means. According to Pitron, more and more consumers around the world would be willing to foot the bill.

The author ends his thesis on a positive note by giving examples of the 'sudden wake-up call taking place in the rare metals industry'.

The ecological transition of our economic activities is critical, not just for saving the planet, but for preserving *life* on the planet — including human life. We can expect hundreds more such challenges to overcome, difficult decisions to be made, scientific breakthroughs to reach, and opinions to support or enlist if we are to accelerate this transition. Meanwhile, the clock is ticking.

Through the focus of his investigation, Guillaume Pitron alerts us to an issue that is vital yet inadequately considered.

Introduction

FOR FOUR THOUSAND YEARS, HUMANS DEPENDED ON FIRE, CAPRICIOUS winds and currents, and manpower and horsepower to roam, build fortresses, and work the land. Energy was a rare and precious resource, movement was slow, economic growth sluggish. Progress came in fits and starts, and history tended to be made one slow step at a time.

Then, from the eighteenth century, humans used the steam engine to power their looms, propel locomotives, and float battleships to reign over the seas. Steam powered the first industrial revolution. This was the world's first energy transition, and one underpinned by the use of an indispensable fuel: a black stone called coal.

In the twentieth century, humans cast aside steam for another innovation: the petrol engine. This technology made vehicles, boats, and tanks more powerful, and paved the way for a new machine — the aeroplane — to conquer the skies. This second industrial revolution was also an energy transition, this time relying on the extraction of another resource: a rock oil called petroleum.

The disruptive effects of fossil fuels on the climate since the turn of the current century have driven humanity to develop new and supposedly cleaner and more efficient inventions — wind turbines, solar panels, electric batteries — that can connect to high-voltage ultra-performance grids. After the steam engine and the internal-combustion engine, these 'green' technologies have shifted us into a third energy and industrial revolution that is changing the world as we know it. Like its two predecessors, this revolution draws on a resource so vital that energy experts, techno-prophets, heads of state, and military strategists already refer to it as The Next Oil of the twenty-first century.

What resource are we talking about?

Most people don't have the slightest idea.

Changing the way we produce and therefore consume energy is humanity's next great adventure. Political leaders, Silicon Valley entrepreneurs, proponents of more moderate consumption, Pope Francis, and environmental groups have urged us to make this change, curbing global warming and saving ourselves from a second flood. Never have empires, religions, and money been so aligned behind a single undertaking.[1] The proof of this — described by former French president François Hollande as the 'first universal agreement in our history' — is neither peace treaty, nor trade deal, nor financial regulation.[2] The Paris agreement that was signed in 2015 following the twenty-first conference of parties to the United Nations Framework Convention on Climate Change (COP 21) is, in fact, an energy treaty.

The technologies we use every day might change, but our primary need for energy will not. Yet, faced with the question of what resource could possibly replace oil and coal as we embrace a new and greener world, no one really knows the answer. Our nineteenth-century ancestors knew the importance of coal, and the enlightened man on the street in the twentieth century was well aware of the need for oil. But today, in the twenty-first century, we are unaware that a more sustainable world is largely dependent on rock-borne substances called rare metals.

Humans have long mined the big names in primary metals: iron, gold, silver, copper, lead, aluminium. But from the 1970s, we turned our sights to the superb magnetic, catalytic, and optical properties of a cluster of lesser-known rare metals found in terrestrial rocks in infinitesimal amounts. Some of the members of this large family sport the most exotic names: rare earths, vanadium, germanium, platinoids, tungsten, antimony, beryllium, fluorine, rhenium, tantalum, niobium, to name but a few. Together, these rare metals form a coherent subset of some thirty raw materials with a shared characteristic: they are often associated with nature's most abundant metals.

As with all elements extracted from nature in Lilliputian quantities, rare metals are concentrates packed with remarkable properties. It is a long and painstaking process, for instance, to distil orange blossom essential oil, but the perfume and therapeutic powers of a single drop of this elixir continue to astound researchers.[3] Producing cocaine deep in the Colombian jungle is no easier feat, yet the psychotropic effects of just one gram of the powder can completely deregulate your central nervous system.[4]

The same applies to the rarest of the rare metals. Eight and a half tonnes of rock need to be purified to produce a kilogram of vanadium; sixteen tonnes for a kilogram of cerium; fifty tonnes for the equivalent in gallium; and a staggering 1,200 tonnes for one miserable kilogram of the rarest of the rare metals: lutecium.[5] (See the periodic table of elements in Appendix 1.) These effectively form the 'primary asset' of the Earth's crust: a concentration of atoms with outstanding properties, fine-tuned by billions of years of geological activity. Once processed industrially, a minute dose of these metals emits a magnetic field that makes it possible to generate more energy than the same quantity of coal or oil. And this is the key to 'green capitalism': the replacement of resources that emit billions of tonnes of carbon dioxide with resources that do not burn and therefore do not generate the slightest gram of it.

Less pollution, and at the same time a lot more energy. Tellingly, one of these elements was given the name 'promethium' by chemist Charles Coryell in the 1940s.[6] His wife, Grace Mary, suggested the name, based on the Greek myth of the Titan Prometheus, who was helped by the goddess Athena to break into the realm of the gods, Olympus, and steal the sacred fire ... to give to humanity.

The name says a great deal about the promethean power that we have acquired by harnessing rare metals. Like demigods, we have carved out a multitude of applications in two fundamental areas of the energy transition: supposedly 'green' technologies and digital technologies. Today, we are assured that the convergence of the two will create a better world. The first examples of this convergence (wind turbines, solar panels, and electric cars) are packed with rare metals to produce decarbonised energy that travels through high-performance electricity grids to enable power savings. Yet these grids are also driven by digital technology that is heavily dependent on these same metals. (See Appendix 11 for the main industrial applications of rare metals.)

Jeremy Rifkin, a leading US theorist of this energy transition and the resulting third industrial revolution, takes this a step further.[7] He writes that the crossover of green technologies and new technologies of information and communication (NTIC) already enables each of us to abundantly and inexpensively generate and share our own 'green' electricity. In other words, the mobile phones, tablets, and computers we use every day have become the key components of a more environmentally friendly economic model. Rifkin's prophecies are so compelling that he now counsels numerous heads of state, and is advising a region in the north of France on how best to implement its new-energy models.[8]

Recent history lends substance to his predictions: in the space of ten years, wind energy has increased seven-fold, and solar power by forty-four. Renewable energy already accounts for 19 per cent of world final energy consumption, with Europe planning to increase its share to 27 per cent by 2030.[9] Even technologies based on combustion engines use these metals to make vehicle and aircraft design lighter, more efficient, and therefore less fossil-fuel-intensive.

Enter the military, which is pursuing its own energy transition. Or *strategic* transition. While generals are unlikely to lose sleep over the carbon emissions of their arsenals, as oil reserves dwindle they will nevertheless have to consider the possibility of war without oil. Back in 2010, a highly influential American think tank instructed the US army to end its reliance on fossil fuels by 2040.[10] How will they do this? By using renewable energy, and by raising legions of electrically powered robots. These remote-controlled weapons, which can be recharged using renewable-energy plants, would be a formidable destructive force and solve the conundrum of getting fuel to the front line.[11] This form of combat is, in fact, already colonising new virtual territories: cyber armies alone could win future conflicts by targeting the enemy's digital infrastructure and altering its telecommunication networks.[12]

Like army generals, we too are engaged in a transition to a connected world in which the way we use digital technology will replace certain resources with nothing but … thin air: clouds, intangible messaging, and online traffic instead of highway traffic. The digitalisation of the economy — we are assured — will drastically reduce our physical footprint on the living world. We stand only to gain from an energy and digital revolution: two technological forces marching hand in hand towards a better world.

Even the face of international relations is changing, as diplomats use rare metals to drive a geopolitical transition. Indeed, the rise of new non-carbon energy, say geopolitical experts, will upend the relationship between oil-producer states and oil-consumer states. It will enable the US to progressively shift its warships from the Straits of Hormuz and Malacca — today's vital oil-transit chokepoints — and rethink its partnership with the Gulf petro-powers. As for the European Union, less reliance on Russian, Qatari, and Saudi Arabian fossil-fuel imports will increase its member states' energy sovereignty.

For all these reasons, the energy transition promises to be positive — although implementing it will be no easy feat so long as we have not seen the last of oil and coal.[13] The world that is taking shape before our eyes nevertheless gives us reason to hope. More modest energy consumption will naturally stave off global tensions around the ownership of fossil-fuel sources, create green jobs in leading industrial sectors, and make Western countries serious energy contenders once again.[14] Irrespective of what Donald Trump thinks, this transition is unstoppable: it involves big money that is pulling in players from all across the economy — including the oil giants.

This energy transition traces its beginnings to Germany in the 1980s,[15] and culminated in Paris in 2015, when 195 nations jointly agreed to accelerate this formidable journey. Their goal: to keep the increase in global warming to below two degrees by the end of this century, mainly by replacing fossil fuels with green energy.

But just as the delegates were about to sign the Paris agreement, a wise old man with pale-blue eyes and a bushy beard, dressed like a hermit descending from the mountain, entered the vast hall of COP 21. With an enigmatic smile on his face, he parted the amassed heads of state. Reaching the podium, he began to speak in a deep and deliberate voice: 'Your intentions are charming, and we can all rejoice in the new world to which you are about to give birth. But you are blind to the perils inherent in your audacity!'

Silence.

Turning to the Western delegates, he continued: 'This transition will cripple entire swathes of your economies, and the most strategic at that. It will plunge hordes of workers into retrenchment, triggering social upheaval that will shake your democratic foundations. Even your military sovereignty will be compromised.'

Now addressing all delegates, he added: 'The energy and digital transition will devastate the environment in untold ways. Ultimately, the environmental price of building this new civilisation is so staggering that there is no guarantee you will succeed.'

He ended with an oracular message: 'Your power has blinded you to the point that you have lost the humility of the sailor before the ocean, the climber before the mountain. You forget that the Earth will always have the final say!'

The wise old man is, of course, a figment of my imagination. But the message is real enough, and crystal clear: the 196 delegations in Le Bourget that day signed the Paris agreement and committed to this Herculean task without considering a few crucial questions.[16] Where and how are we going to procure the rare metals without which this treaty will fail? Will there be winners and losers on the new chessboard of rare metals, as there were for coal and oil? And what will be the economic, social, and environmental cost of securing their supply?[17]

For eight years and across a dozen countries, I researched these new rare substances that are already upending our world. I ventured deep into mines in the tropics of Asia, eavesdropped on deputies in the corridors of the French National Assembly, flew over the deserts of California in a light aircraft, bowed before the queen of a tribal community in southern Africa, travelled to the 'cancer villages' of Inner Mongolia, and blew the dust off old parchments in venerable London institutions.

Across four continents, men and women involved in the opaque and underground world of rare metals shared with me a very different and far darker tale of the energy and digital transition. By their account, the emergence of these new substances in the wake of fossil fuels has not done us or the planet any of the favours we would expect from a supposedly greener, friendlier, and more insightful world — far from it.

Great Britain dominated the nineteenth century, thanks to its hegemony over global coal production. Many of the events of the twentieth century can be seen through the lens of US and Saudi Arabian control over oil production and supply routes. In the twenty-first century, one state is in the process of dominating the export and consumption of rare metals. That state is China.

Consider this economic and industrial observation: by committing to the energy transition, we have flung ourselves headlong into the jaws of the Chinese dragon. Arguably, the Middle Kingdom holds a near monopoly over a profusion of rare metals without which low-carbon and digital energies — the very foundations of the energy transition — cannot exist. And, as I will address later in this book, China has used barely credible chicanery to position itself as the sole supplier of the most strategic of the rare metals. Known as 'rare earths', they are difficult to substitute, and the vast majority of industrial groups cannot do without them.[18] (See Appendix 12 for the main industrial applications of rare earths.)

And so the West has placed the fate of its green and digital technologies — the cream of its industries of the future — in the hands of just one nation, while China is nurturing its own technologies and playing hardball with the rest of the world by putting a cap on the export of those resources. The result: serious economic and social consequences for the rest of the world.

Next, an ecological observation: our quest for a more ecological growth model has resulted in intensified mining of the Earth's crust to extract the core ingredient — rare metals — with an environmental impact that could prove far more severe than that of oil extraction. Changing our energy model already means doubling rare metal production approximately every fifteen years. At this rate, over the next thirty years we will need to mine more mineral ores than humans have extracted over the last 70,000 years. But the shortages already looming on the horizon could burst the bubble of Jeremy Rifkin, green-tech industrialists, and Pope Francis, and prove our hermit right.

The third observation relates to geopolitics and the military. The continued existence of the most sophisticated Western military equipment (robots, cyberweapons, and fighter planes, including the US's supreme F-35 stealth jet) also partly depends on China's goodwill. This has US intelligence leaders concerned, especially as one high-ranking US army officer states that 'only war can now stop Beijing controlling the South China Sea'.[19]

This latest scramble for resources is already heightening tensions over ownership of the most abundant deposits, sparking territorial conflicts in peaceful backwaters apparently of interest to no one. Fuelling this thirst for rare metals are a burgeoning global population set to reach 8.5 billion by 2030, the boom in new modes of high-tech consumption, and the growing convergence of Western and emerging economies.[20]

By seeking to break free from fossil fuels and turn an old order into a new world, we are in fact setting ourselves up for a new and more potent dependence. Robotics, artificial intelligence, digital healthcare, cybersecurity, medical biotechnology, connected objects, nanoelectronics, driverless cars ... the most strategic sectors of the economies of the future, all the technologies that will exponentially increase our computing capacity and modernise how we consume energy, our daily routines, and even our most significant collective choices will depend entirely on rare metals. These resources will provide the fundamental building blocks of the twenty-first century. Yet our addiction is already pointing to a future so far unpredicted. We thought we could free ourselves from the shortages, tensions, and crises created by our appetite for oil and coal. Instead, we are replacing these with an era of new and unprecedented shortages, tensions, and crises.

From tea to black oil, nutmeg to tulips, saltpetre to coal, commodities have been a backdrop to every major exploration, empire, and war, often altering the course of history.[21] Today, rare metals are changing the world. Not only are they polluting the environment, but they are jeopardising economic stability and global security. In the twenty-first century alone, their use has consolidated China's supremacy and accelerated the weakening of the West that began at the turn of the millennium.

But the rare metals war is far from lost. China has made some colossal errors; the West can respond; and the technical progress we have yet to make is bound to transform how we generate wealth and energy.

Until then, this book recounts the dark side of the story of the world that awaits us. It is an undercover tale of a technological odyssey that has promised so much, and a look behind the scenes of our lavish and ambitious quest that involves risks as formidable as those it sets out to resolve.

CHAPTER ONE

The rare metals curse

'WHAT DO YOU WANT? YOU HAVE NO BUSINESS HERE!' A MAN IN HIS forties pulls up to us in a black Audi, and stares at us menacingly. He is joined by a companion, equally menacing, and soon another man on a motorbike also pulls up. 'You need to leave, it's dangerous. We don't want any trouble!' The three men start to lose their cool; the tension is palpable. 'Get lost!' shouts the man in the Audi. He can tell that I am trying to buy time by my furtive glances at the tent erected incongruously on the hillside.

'People still work here,' whispered Wang Jing, a former miner and my scout. 'I was sure they'd closed these quarries ages ago!' The scatter of new equipment and evacuation pipes in the area confirm my doubts. Two hundred metres away, the telltale tent stands above the tailing ponds and disembowelled rocky landscape. There can be no doubt that rare metal–refining activities are taking place at this camp. Where are they extracting the minerals from? 'From the mines all around us, but also from the enormous illegal quarries flanking the hill', Wang Jing explains.

Two days earlier that July in 2016, I had landed at the tiny airport of Ganzhou, a town in the Chinese province of Jiangxi some 700 kilometres south of Beijing. From there, I drove due south for hours on battered roads, hemmed in by row upon row of rice paddies, to reach the mines. The last few dozen kilometres took me along little more than ribbons of asphalt, weaving between rickshaws and trailers laden with rubble, and women wearing the traditional *cǎo mào* conical hat. Out of the foothills of the Nan Kang mountains burst lotus forests and palm trees — a lush and abundant organic kingdom of stifling foliage pushing into the blue sky.

I was also in the biggest rare metals mining area on the planet.

Rare metals: a definition

When it comes to raw materials, nature can be surprisingly generous or deeply parsimonious. Alongside popular species such as the poplar and pine are rare trees like the hairy quandong in Australia or the ghost orchid in the UK. Tulips may very well overflow the fields of Holland, but other flowers, such as the butterfly orchid, barely make an appearance in flower shops in the Netherlands. In these parts, the skies abound with birds like the mallard — much to the delight of hunters across Western Europe. Then there are the more discreet, rarely sighted birds like the California condor in North America.

Similarly, abundant metals like iron, copper, zinc, aluminium, and lead coexist with a family of some thirty rare metals.[1] The lists published by the United States Geological Survey, an agency of the United States Department of the Interior, and by the European

Commission are an education in themselves: light and heavy rare-earth elements, germanium, tungsten, antimony, niobium, beryllium, gallium, cobalt, vanadium, tantalum, and other rare metals.[2]

They share the following traits:

- They are associated with abundant metals found in the Earth's crust, but in minute proportions. For instance, there is 1,200 times less neodymium and up to 2,650 times less gallium than there is iron.

- Naturally, this is reflected in the markets. Rare metal production is on a minute scale, and is overlooked by mainstream media: every year, 160,000 tonnes of rare-earth metals are produced — 15,000 times less than annual iron production of two billion tonnes. Likewise, 600 tonnes of gallium are produced annually, which is 25,000 times less than the 15 million tonnes of annual copper production. (See the European Commission's list of 'critical' metals for the EU in Appendix 13.)

- This makes these rare metals expensive: a kilogram of gallium is worth around US$150 — that is, it is 9,000 times more expensive than iron. Germanium is ten times more expensive than gallium.

- They also possess the exceptional properties demanded by new — and especially 'green' — technology manufacturers working to reduce our carbon footprint on the environment.[3]

Rare metals: drivers of new energies

Since the dawn of time, humans have sought to transform sources of natural energy (such as wind, thermal, and solar) into mechanical energy. Take the windmill, for instance. Its vanes and rotor are driven by wind energy to actuate a mechanical mill that then crushes olives or grain. In the steam engine, thermal energy transported by steam from water is converted, using pistons, into mechanical energy powerful enough to drive a locomotive. Thermal energy is also generated by the combustion of fuel to drive the pistons of a vehicle and set it in motion. In essence, we have been making movement-generating machines for centuries.[4] The more we multiply the possibilities of movement, the more we can travel and trade, entrust new tasks to machines and robots, and make productivity gains — and therefore greater profits.

Energy needs to be both abundant and inexpensive to ensure that machines run efficiently — a challenge we must overcome to satisfy our economic-growth ambitions. Thus, for almost three centuries we have been working tirelessly at developing new engines with increasingly impressive power-to-weight ratios: the more compact and less resource-intensive they are, the greater their mechanical energy output.

Enter rare metals. While mineralogists have known of their existence since the eighteenth century, they garnered little interest while their industrial applications remained undiscovered. But from the 1970s, humans began to exploit the exceptional magnetic properties of some of these metals to make super magnets.[5]

An electrical charge coming into contact with the magnetic field of a magnet generates a force that creates movement. The smallest of these magnets is barely the size of a pinhead, while the biggest magnet ever designed is four metres high, weighs 132 tonnes, and is located at the Saclay Nuclear Research Centre in the Paris region.[6] Irrespective of their size, magnets are now — to a vast majority of electric engines — what pistons have been to steam and internal-combustion engines. Magnets have made it possible to manufacture billions of engines, both big and small, capable of executing certain repetitive movements in our stead — whether it be running a motorbike, powering a train, making an electric toothbrush or mobile phone vibrate, operating an electric window, or launching an elevator to the top of the tallest skyscraper.

Without realising it, our societies have become completely magnetised. To say that the world would be significantly slower without magnets containing rare metals is not an understatement.[7] Remember that the next time you stop to admire your holiday magnets on the fridge!

The technological revolution behind the energy shift

Electric engines did more than make humanity infinitely more prosperous; they made the energy transition a plausible hypothesis. Thanks to them, we have discovered our ability to maximise movement — and therefore wealth — without the use of coal and oil. It is not surprising that electric engines will soon replace conventional engines. Electric engines are already being used to propel ships, send the Solar Impulse aircraft around the world, launch space probes and satellites, and put enough electric cars on the roads to disrupt the automotive market.[8]

Of course, these engines run off electric batteries that create the electricity needed to activate the magnets. The difference, however, is that with rare metals it is possible to generate clean energy: they cause the rotors of certain wind turbines[9] to turn and convert the sun's rays into electricity using solar panels.[10] Because they remove pollution from most of the energy cycle — from manufacture to end use — we can safely envisage a world without nuclear, oil-fired, or coal-fired power plants.

But that is merely scratching the surface of rare metals, for they possess a wealth of other chemical, catalytic, and optical properties that make them indispensable to myriad green technologies.[11] An entire book could be written on the details of their characteristics alone. They make it possible to trap car-exhaust fumes in catalytic converters, ignite energy-efficient light bulbs,[12] and design new, lighter, and hardier industrial equipment, improving the energy efficiency of cars and planes. Two thousand years ago, the Hebrews were able to cross the Sinai desert surviving on manna, the providential food sent from heaven. Today, another godsend — this time from underground — has been laid out at the ecological banquet, for there is a rare metal for every green application. Surely there is a green guardian angel watching over us.

Most surprising is how these metals have become indispensable to new information and communication technologies for their semiconducting properties that regulate the flow of electricity in digital devices. And the once-distinct functions of green and digital technologies are beginning to converge. Indeed, increasingly sophisticated software and algorithms used in 'smart grids' make it possible to regulate fluctuations in the flow of electricity between producers and consumers. This is precisely what the 80 million smart meters already installed in the US are doing. In the smart cities of tomorrow, which will combine green and digital technologies, we will save up to 65 per cent of the electricity we use today, thanks to sensor-embedded streets that adjust the lighting to foot traffic, while weather-prediction software makes solar panels 30 per cent more efficient.

Thus, digitalisation and the energy transition are co-dependent. Digital technology advances and enhances the impacts of green tech. Their combination is ushering in an era of energy abundance, stimulating new industries, and has already created 10 million jobs worldwide.[13] This is a boon not lost on political leaders: to help these new markets take off, Europe is now urging its member states to reduce their carbon dioxide emissions by 40 per cent (in relation to 1990 levels) by 2030, and to increase to 27 per cent the renewable-energy share of their energy consumption.

But why stop there? A 2015 report by the Royal Society of Chemistry confirmed it was economically and technically feasible for the US to rely only on renewable-energy sources by 2050.[14] In 2019, Democratic representative Alexandria Ocasio-Cortez defended the very same objective under the 'Green New Deal'.[15]

The acceleration of rare metal consumption

This technological diversification has multiplied the types of metals that humanity uses. Between the ages of antiquity and the Renaissance, human beings consumed no more than seven metals;[16] this increased to a dozen metals over the twentieth century; to twenty from the 1970s onwards; and then to almost all eighty-six metals on

Mendeleev's periodic table of elements. (See Appendix 1.)

Our appetite for metals boomed — and it didn't stop there. On the one hand, consumption of the three main sources of energy currently used in the world (coal, oil, and gas) tends to stabilise, decrease, or, at best, moderately increase.[17] On the other hand, the potential demand for rare metals is exponential. We are already consuming over two billion tonnes of metals every year — the equivalent of more than 500 Eiffel Towers a day.[18] (See Appendix 2 to see the trends in world primary metal production.) By 2035, demand is expected to double for germanium; quadruple for tantalum; and quintuple for palladium. The scandium market could increase nine-fold, and the cobalt market by a factor of 24.[19] There is going to be a scramble for these resources, for the resilience of capitalism relies increasingly on the emergence of green and digital technologies. The market will become less and less dependent on the fuels of the last two industrial revolutions, and will increasingly rely on the metals that are driving the impending transition.

The US Geological Survey and the European Commission agency in charge of raw materials have produced a map of the world's rare metal production areas. It shows that South Africa is a major producer of platinum and rhodium; Russia of palladium; the US of beryllium; Brazil of niobium; Turkey of borates; Rwanda of tantalum; and the Democratic Republic of Congo (DRC) of cobalt. Yet most of these metals come from Chinese mines. This is the case for antimony, germanium, indium, gallium, bismuth, tungsten, and, above all, the supreme 'green' metals, whose staggering electromagnetic, optical, catalytic, and chemical properties surpass all others in performance and fame: the rare-earth metals.

They form a family of seventeen elements, featuring exotic names like scandium, yttrium, lanthanum, cerium, praseodymium, neodymium, samarium, europium, gadolinium, terbium, dysprosium, holmium, erbium, thulium, ytterbium, lutetium, and promethium. (See Appendix 3 for a map of rare-mineral-producing countries.)

Rare earths, the black market, and environmental disasters

The biggest quantity of rare earths is extracted from the bowels of Jiangxi, in the heart of tropical China, which is where our story begins.

Wang Jing knows this better than anyone. I had met the fresh-faced 24-year-old, with smiling eyes under his mop of hair, in the village of Xing Quang. He knows this strip of mountain like the back of his hand, and had little difficulty guiding me there. In fact, he spent years working at this illegal mine concealed by a copse of eucalyptus trees. He tells me how he would chip away at red-tinted rock and crush prodigious rubble aggregates alongside other miners, both men and women.

Like a human anthill, the mountain was mined twenty-four hours a day, seven days a week. Miners were paid a few hundred euros a month, and slept on ground plundered by picks and excavators. At this frenetic pace, hundreds of thousands of tonnes of minerals were extracted from the mountain. But four years earlier, the Chinese authorities had banned these activities, and illegal miners were slapped with heavy fines. Entire stocks of metals promised to foreign markets were seized at the port of Canton, some hundreds of kilometres south, and dozens of traffickers were thrown into prison.

Despite this, the more determined and needy miners have entrenched themselves in the folds of the mountain's most inhospitable terrains. They prosper in secret, and are said to make payoffs to the local police. Their activities are feeding a colossal Chinese black market for minerals that, once processed, are exported worldwide.

These are the activities I have come to see. And the three illegal miners standing in our way know it. The motorcyclist threatens me again. I move away from the tent; clearly, I won't get to see what I came for: proof of the staggering pollution created by rare-earth mining.

'It's poison,' says Wang Jing. 'The chemicals used for refining the minerals were poured straight into the ground.' The sulphuric and hydrochloric acids would pollute the nearby stream to the point that 'it was impossible for any plant to grow'. Because the closest housing lies far from the Yaxi mountains, there was no visible impact on residents. But elsewhere, housing was much closer, he said.

The 10,000 or so mines spread across China have played a big role in destroying the country's environment.[20] Pollution damage by the coal-mining industry is well documented. But barely reported is the fact that mining rare metals also produces pollution, and to such an extent that China has stopped counting contamination events. In 2006, some sixty companies producing indium — a rare metal used in the manufacture of certain solar-panel technologies — released tonnes of chemicals into the Xiang River in Hunan, jeopardising the meridional province's drinking water and the health of its residents.[21] In 2011, journalists reported on the damage to the ecosystems of the Ting River in the seaside province of Fujian, due to the operation of a mine rich in gallium — an up-and-coming metal for the manufacture of energy-efficient light bulbs.[22] And in Ganzhou, where I landed, the local press recently reported that the toxic waste dumps created by a mining company producing tungsten — a critical metal for wind-turbine blades — had obstructed and polluted many tributaries of the Yangtze River.

A Chinese journalist reporting anonymously describes the working conditions — reminiscent of a bygone era — at the graphite mines of Shandong, in eastern China. In the processing plants rising out of dark, uprooted mounds of the Earth's crust, '[M]en and women, wearing no more than basic face masks, work in areas thick with black particles and acid fumes. It's hell.' To complete the picture are toxic pits of chemical discharges from the plants, fields of poisoned corn, acid rain, and more. 'Local authorities tried to police environmental offences,' says the journalist, 'but the pressure from automobile manufacturers was too great.'

Dirty metals for a greener world

The assertion that producing the metals we need for a cleaner world is in fact a polluting process seems incomprehensible at first. Which is understandable: most consumers have forgotten what they learned in their high-school natural sciences, physics, and chemistry classes. Let's refresh our memories.

No need to dust off the chalkboard; a trip to the closest bakery will do. Everyone knows the ingredients of a loaf of bread: a good portion of flour, water, a bit of yeast, and a pinch of salt. It's not that different from a rock of a similar size taken out of a mine: its ingredients comprise several minerals mixed together.

In this metaphor, the flour represents the rock that ends up on the rubble heap. The water is where it starts to get interesting: all things being equal, it represents iron — a mineral found in abundance in the Earth's crust. Next is the yeast, making up a much smaller portion of the mix, which represents nickel — a metalloid rarer than iron. That leaves the pinch of salt: our rare metals. Their concentration in the Earth's crust is as minute and imperceptible as the pinch of salt sprinkled into our bread dough.

But rock is composed of minerals aggregated over billions of years, and the rare metals are therefore completely incorporated in the rock — just like the salt when it is kneaded and baked into the dough. You would think that trying to extract it would be practically impossible, yet decades of research have developed the chemical processes to do just that. And the Chinese 'sorcerer's apprentices' deep in the mines of Jiangxi province and elsewhere are managing to achieve this: to extract rare metals from rock.

For a process known as 'refining', there is nothing refined about it. It involves crushing rock, and then using a concoction of chemical reagents such as sulphuric and nitric acid. 'It's a long and highly repetitive process,' explains a French specialist. 'It takes loads of different procedures to obtain a rare-earth concentrate close to 100 per cent purity.'

That's not all: purifying a single tonne of rare earths requires using at least 200 cubic metres of water, which then becomes saturated with acids and heavy metals.[23] Will this water go through a water-treatment plant before it is released into rivers, soils, and ground water? Very rarely. The Chinese could have opted for clean mining, but chose not to. From one end of the rare metals production line to the other, virtually nothing in China is done according to the most basic ecological and health standards. So as rare metals have become ubiquitous in green and digital technologies, the exceedingly toxic sludge they produce has been contaminating water, soil, the atmosphere, and the flames of blast furnaces — representing the four elements essential to life. The result is that producing rare metals has become one of the most polluting — and secretive — industries in China. But that won't stop me from taking a closer look.

My next stop was Hanjiang, a few dozen kilometres from the rare-earth mines I surveyed with Wang Jing. It is a hamlet located close to another of these mines. But 90 per cent of its inhabitants had fled the jumble of stone houses and their dark-tiled roofs. The residents complained that because of the rampant mining activities, '[n]othing we plant grows anymore. Our rice paddies have become infertile!' Those who have refused to leave have accepted their fate. 'What can we do?' asked an old man, overwhelmed by the thick, cloying air. 'There's no point even complaining about it.' Do the local authorities know about the pollution? 'Of course they do! Even you would have guessed without anyone telling you!'

The heavy toll on health

This is nothing compared to what awaits me 2,000 kilometres north in Baotou, the capital of the autonomous region of Inner Mongolia, which I first visited in 2011. It's a city well known to all rare metal hunters for the simple reason that it is the biggest rare-earth production site on the planet, far surpassing Jiangxi province. There I saw convoys of trailer trucks laden with gravel trundle down the dusty roads of the city and surrounding countryside. The hundreds of thousands of tonnes of rare earths extracted annually by the mining giant Baogang — responsible for 75 per cent of global production — contributes to the prosperity of the city and its three million or so inhabitants.

It must be said that I find Baotou quite pleasant, with its flurry of Chinese flags waving from the roofs of buildings, and swarms of bicycles zipping between the city and its industrial areas, while the troubled waters of Asia's second-longest river, the Yellow River, caress the city's edges. At the entrances of the city's parks are hundreds of posters depicting a couple and their child against a green, pristine background bearing the slogan: *Building a clean city for our country*. It's postcard perfect.

It is impossible, however, to get anywhere near the Baogang mines, some 100 kilometres from the city centre. Having already been marched to the police station by a pair of overzealous police officers, I was in no hurry to go back. But my Chinese fixer reckoned that by going just a few dozen kilometres west of the city, I could catch a glimpse of the industry's secretive activities.

Past the suburbs of Baotou, below a quadruple carriageway, a lonely path led me to a cement embankment bristling with pylons, each one equipped with a security camera to watch for intruders. This is how I reached the Weikuang Dam — an artificial lake into which metallic intestines regurgitate torrents of black water from the nearby refineries. I was looking at 10 square kilometres of toxic effluent, which occasionally flows over into the Yellow River.

This is also the beating heart of the energy and digital transition.

I was left speechless during the hour I spent observing this immense, disintegrating lunar landscape. Wang Jing and I decided to get moving before the security cameras alerted the police to our presence.

A few minutes later, we arrived in Dalahai, on another side of the artificial lake. In this village of redbrick houses, where the thorium concentration in the soil is in some places thirty-six times higher than in Baotou, the thousands of villagers still living there breathe, drink, and eat the toxic discharge of the reservoir. We met one of them, Li Xinxia. With her striking features and wistful eyes, the 54-year-old woman knew this was a touchy subject, but confided in me anyway: 'There are a lot of sick people here. Cancer, strokes, high blood pressure … almost all of us are affected. We are in a grave situation. They did some tests here, and our village was nicknamed "the cancer village". We know the air we are breathing is toxic and that we don't have that much longer to live.'

Is there any way out for Li Xinxia and her loved ones? The provincial authorities did offer the villagers 60,000 yuan per mu of land (around US$9,000 for 666 square metres) to relocate to high-rises built in a neighbouring town. While it was a handsome sum in a rural area where the average annual income is around US$1,300, it was not warmly greeted by the farmers. The apartments were prohibitively priced for people who could no longer live off land that has become infertile.

Rare earths have cost the community dearly. The hair of young men barely thirty years of age has suddenly turned white. Children grow up without developing any teeth. In 2010, the Chinese press reported that sixty-six Dalahai residents had died of cancer.

I return to Baotou in spring 2019. The town has expanded considerably, and its suburbs now encroach on a succession of huge rare-earth refineries. According to a few people we questioned discreetly on the site's perimeter, the industrial group had destroyed everything before it extended its operations — starting with Dalahai, where the 'cancer villages' were razed and their inhabitants compensated to move. All that remains are pieces of bricks that men and women from neighbouring villages come and clear at the end of the afternoon when the heat is less oppressive.

Weikuang Dam still lies between large man-made embankments, and is fed incessantly by factory effluents. 'The reservoir is massive. If you look at it from the roof of a building, you'll understand. It's so big you'd think it was the sea!' The presence of Chinese security nearby does not stop Gao Xia from giving us her testimony. The 48-year-old villager has been rehoused with her husband in a gloomy high-rise estate overlooking the devastated landscape. Eight years after I first covered the story, the same causes seemed to be producing the same effects. By Gao Xia's account, it is a disaster area. The water in the rivers 'is whitish-green and sometimes red', and the land produces corn and buckwheat with great difficulty, while cancer continues to affect local populations. After a life of living off the land, Gao Xia is condemned to eke out a living doing odd jobs here and there. She speaks of her 'bitterness' at being powerless against the rare-earth companies 'that have polluted our environment'.

'The Chinese people have sacrificed their environment to supply the entire planet with rare earths,' Vivian Wu, a recognised Chinese expert in rare metals, tells us. 'Ultimately, the price of developing our industry is just too high.'

How could Beijing have allowed this disaster to happen?

Playing catch-up at the risk of anarchy

To answer this question, we need to go back in time. The nineteenth and twentieth centuries were times of decline and humiliation for the Middle Kingdom. At the death of the Qianlong emperor — the 'Chinese Louis XIV' — in 1799, China was the global powerhouse. The empire's borders reached the furthermost bounds of Mongolia, Tibet, and Burma. More clement temperatures and bountiful harvests led to a population boom, and at the zenith of the Qing dynasty, the political system was stable and the country's economic production represented one-third of global gross domestic product (GDP). Middle Kingdom mania extended as far as Europe: the French writer, historian, and philosopher Voltaire waxed lyrical on the merits of Manchurian autocracy, *chinoiserie* was all the rage, and the English discovered their love of tea.

But soon the edifice crumbled, followed by one disaster after another: opium wars,[24] unfair treaties, humiliation at the Treaty of Versailles in 1919 (despite China being one of the victors of the First World War),[25] the failures of the Kuomintang party,[26] and the devastating effects of Maoism. At the death of Mao Zedong in 1976, China's position in the world economy had diminished tenfold compared to the end of the eighteenth century. The country had been ravaged by civil wars, and the survivors of the bloody Cultural Revolution (which killed millions) were subjected to ghastly brainwashing.

But the Chinese are resilient; their hunger to recover their lost prestige is insatiable. After all, between the year 960 and today, China was the leading global power for close to nine centuries. The Middle Kingdom had to take back its place — at all costs.

Obsessed by the idea of erasing the failings of the nineteenth and twentieth centuries as quickly as possible, China has been racing at reckless speed to achieve in three decades the economic progress that took the West three centuries to accomplish. In 1976, the Communist Party, under the leadership of Deng Xiaoping, opened the country to capitalism and global trade. Its policy of combining economic and environmental dumping in the form of below-market-value prices and lax environmental rules gave China a competitive advantage over Western countries, making it the factory of the world and the West's official supplier of low-cost goods. Lastly, and most importantly, Beijing became the primary producer of all the minerals that the world needs to support its economic growth. Today, China is the leading producer of twenty-eight mineral resources that are vital to our economies, often representing over 50 per cent of global production.[27] It also produces at least 15 per cent of all mineral resources, other than platinum and nickel.[28] (See Appendix 4 on China's relative share of global mining and metallurgy production in 2011.)

The downside of this spectacular success? Little attention has been given to the environmental impact of these economic choices. Industry has been left to pollute the atmospheres of major cities, to contaminate the soil with heavy metals, and to dump its mining waste in most rivers with impunity. Under the growth measures in place, anything goes. In other words, the Chinese have made a real mess of it.

The environmental cost is exorbitant, inhumane, and outrageous.[29] China is the biggest emitter of greenhouse gases (producing 28 per cent of global greenhouse-gas emissions in 2015), and the alarming figures coming out of the country are multiplying. Ten per cent of its arable land is contaminated by heavy metals, and 80 per cent of its ground water is unfit for consumption. Only five of the 500 biggest cities in China meet international standards for air quality, and there are 1.6 million deaths per year due to air pollution alone.[30] In the words of Chinese environmental activist Ma Jun, whom I met in Beijing: 'This was a monumental error.'

The scourge of rare metals gone global

The pollution caused by rare metals is not limited to China. It concerns all producing countries, such as the Democratic Republic of Congo, which supplies more than half the planet's cobalt. This resource — indispensable to the lithium-ion batteries used in electric vehicles — is mined under conditions straight out of the Middle Ages. One hundred thousand miners equipped with spades and picks dig into the earth to find the mineral, especially in the southern region of Lualaba. Given the DRC government's inability to regulate the country's mining activities, the pollution of surrounding rivers and turmoil in the ecosystems are legion. Research by Congolese doctors has found that the cobalt concentration in the urine of the local communities living near the mines of Lubumbashi, in Katanga province, is up to 43 times higher than a control sample.[31]

We see the same in Kazakhstan, a central Asian country that produces 14 per cent of the world's chrome — prized by the aerospace industry for the manufacture of superalloys that improve the energy performance of aircraft.[32] In 2015, researchers from South Kazakhstan State University discovered that chrome mining was responsible for the colossal pollution of the Syr Darya, the longest river in Central Asia. Its water had become completely unfit for consumption by the hundreds of thousands of inhabitants, who are now even advised against using it for their crops.[33]

Latin America has already started to experience similar problems with lithium mining — a white metal lying below the salt flats of Bolivia, Chile, and Argentine. It is considered critical by the US, and demand is expected to soar on the back of the electric car boom that has jacked up its global production. Naturally, Argentina has its sights set on becoming the giant of lithium, and between now and 2025 the country has the capacity to produce up to 165,000 tonnes a year, or 45 per cent of global demand, provided it can get foreign investors on board.[34]

In May 2017, all the rare metal exploration, mining, and refining companies operating in Latin America met on the banks of Rio Plata near Buenos Aires for the Arminera international mining trade fair. Amid the excavators, skips, light towers, and other waste-water treatment equipment on display, Daniel Meilán, Argentina's mining secretary, boasted about the 'dozens of prospecting activities for lithium deposits in progress' in the country, and promised a mining sector that would be responsible and compliant with international ecological standards. Before popping the champagne, and to the applause of those present, all the Argentinian industry players were invited to sign an ethics charter.

At the same time, some thirty Greenpeace activists had blocked the entrance to the trade fair, brandishing banners calling out the lies of the mining industry. 'Everything they say is pure greenwashing,' said one of its members, Gonzalo Strano. 'There's no such thing as sustainable mining. Not only does it dig out the ground by definition, it uses chemicals and massive amounts of water, which is a problem.'

The mining sector in Latin America has a sulphurous reputation. From Mexico to Chile, from Colombia to Peru, the last few years have seen growing opposition from local communities. Most emblematic of this deep distrust is the Pascua Lama gold and silver mine, operated by Canadian mining group Barrick Gold, in the north of Santiago, Chile. Extracting at this site would have involved destroying the glacier concealing the orebody — a prospect that had local residents up in arms, forcing Barrick Gold to shut down its activities in 2013.[35]

The Pascua Lama example inspired the entire Latin American mining sector. Large-scale lithium mining now sparks environmental activism. As with any mining activity, it requires staggering volumes of water, diminishing the resources available to local communities living on water-scarce salt flats. Already, the communities of the Hombre Muerto salt flat in Argentina blame lithium mining for contaminating their streams.[36]

Extracting minerals from the ground is an inherently dirty operation. The way it has been carried out so irresponsibly and unethically in the most active mining countries casts doubt on the virtuous vision of the energy and digital transition. A recent report by the Blacksmith Institute identifies the mining industry as the second-most-polluting industry in the world, behind lead-battery recycling, and ahead of the dye industry, industrial dumpsites, and tanneries.[37] It has moved up one rung since the 2013 rankings, in which the much-maligned petrochemical industry doesn't even crack the top ten. Given China's dominant role in the global supply of rare metals, we cannot accurately assess the progress made in combating global warming without properly accounting for Beijing's ecological performance. Which is catastrophic, to say the least.

The message of this overview of the environmental impacts of extracting rare metals from the Earth is clear: we need to be far more sceptical about how green technologies are manufactured. Before they are even brought into service, the solar panel, wind turbine, electric car, or energy-efficient light bulb bear the 'original sin' of its deplorable energy and environmental footprint. We should be measuring the ecological cost of the entire lifecycle of green technologies — a cost that has been precisely calculated.

CHAPTER TWO

The dark side of green and digital technologies

THE TECHNOLOGIES THAT WE DELIGHT IN CALLING 'GREEN' MAY NOT be as green as we think. They could even have an enormously negative impact on the environment, as we learned in Toronto in the spring of 2016.

In the heart of the city's financial district, everyone who's anyone in the North American mining industry — mining companies, experts, public authorities, venture capitalists, consultancies, and academics — gathered in the lavish setting of a grand hotel for a conference on the rare metals 'gold rush'.[1] On the agenda: investments, cashflow, gross margins, capital raising, cost structuring, market capitalisation, and average annual output. With the International Energy Agency expecting the share of renewables in global electricity output to increase from 26 per cent in 2018 to 45 per cent in 2040, the green-tech growth outlook looks bright indeed.[2]

But in the midst of this predominantly male and well-heeled crowd were two people about to throw a spanner in the mining works.

Green tech's heavy toll on the environment

The first was Canadian Bernard Tourillon, director of Uragold — a company that makes equipment for the solar industry. Tourillon has painstakingly calculated the ecological impact of photovoltaic panels: on account of their silicon content, producing just one panel generates as much as 70 kilograms of carbon dioxide. The 23 per cent annual increase in the number of solar panels over the next few years will increase their power-generation capacity by 10 gigawatts every year — but will also generate 2.7 billion tonnes of carbon emissions into the atmosphere, or as much pollution as 600,000 vehicles on the road in a year.[3]

Thermal solar panels make an even bigger splash, with some consuming as much as 3,500 litres of water per megawatt hour.[4] That's 50 per cent more water than a coal-fired plant.[5] More problematic still is the fact that solar farms are more often than not located in water-scarce areas.

The second party pooper was John Petersen, a Texan lawyer with an extensive career in the electric battery industry. Based on number crunching, countless academic papers, and his own research, Peterson reached an astonishing conclusion.

Rewind to 2012, when researchers at the University of California Los Angeles (UCLA) compared the carbon impact of a conventional fuel-driven car against that of an electric car.[6] Their first finding was that the production of the supposedly more energy-efficient electric car requires far more energy than the production of the conventional car. This is mostly on account of the electric car's very heavy lithium-ion battery. The battery of prominent US company Tesla's Model S, for instance, weighs in at 544 kilograms, or 25 per cent of the car's weight. (See Appendix 5 for an overview of the rare metals contained in an electric vehicle.)

Then there's the composition of the lithium-ion battery: 80 per cent nickel, 15 per cent cobalt, 5 per cent aluminium, as well as lithium, copper, manganese, steel, and graphite.[7] By now, we are familiar with the conditions in which these minerals are extracted in China, Kazakhstan, and the Democratic Republic of the Congo. But we also need to consider how these minerals are refined, not to mention the logistics of their transportation and assembly. The UCLA researchers reached the conclusion that industrialising electric vehicles is three to four times more energy-intensive than industrialising conventional cars.

Looking at its end-to-end lifecycle, however, the electric vehicle has the undeniable advantage of not requiring petrol. This brings down its carbon dioxide emissions to 32 tonnes from factory to scrapyard, compared with almost double that for a conventional vehicle.

The caveat of this research is that it was conducted on a medium-sized electric-vehicle battery with a 120-kilometre range in a market that is growing so fast that none of the cars being rolled out today have a range below 300 kilometres. According to Petersen, a battery that is powerful enough to drive a vehicle for 300 kilometres emits twice as much carbon as production-phase emissions — a figure we can then *triple* for batteries with a 500-kilometre range.

Therefore, over its entire lifecycle, an electric car may produce as much as three-quarters of the carbon emissions produced by a petrol car. And the more powerful the electric cars are, the more energy they need for their production, potentially increasing greenhouse gases. In the meantime, Tesla has announced that its Model S vehicles will now be equipped with 600-kilometre-range batteries,[8] and its CEO, Elon Musk, has announced the imminent arrival of 800-kilometre-range batteries.[9]

John Petersen's conclusion? 'Electric vehicles may be technically possible, but their production will never be environmentally sustainable.'[10] This concurs with similar research conducted along the same lines. The 2016 report by the French Environment & Energy Management Agency (ADEME) finds: 'he energy consumption of an electric vehicle [EV] over its entire lifecycle is, on the whole, similar to that of a diesel vehicle.'[11] The report also finds that its environmental impact is 'on a par with [that of] the petrol car'. In fact, an EV might even emit more carbon dioxide than it consumes if the electricity it uses comes predominantly from coal-fired plants, as is the case in countries such as China, Australia, India, Taiwan, and South Africa. These are precisely the conclusions of research published in the journal *Nature Energy* in 2018: if everyone in China were to rush to a fast-charge station during a peak in electricity consumption — when only thermal power plants are able to meet most of this demand — an electric car in China might generate, over its entire life cycle, more carbon dioxide than a conventional car.[12]

This leaves us with a list of unanswered questions. Is the replacement impact of EV batteries, which have a short lifespan, being taken into account? Do we know the precise ecological cost of all the electronics and connected components packed into these electric vehicles? What about the environmental impact of recycling these mostly still-new vehicles in the future? And how much energy will it take to build the necessary electric grids and plants to meet these new needs?[13]

Ultimately, as was quietly admitted to me by a US rare metals expert in Toronto that day, 'No one in the green-energy business is going to communicate on these points. Besides, don't we all want to believe that we're making things better rather than making things worse?'

Tangible invisibility

It doesn't stop there. We know that green technologies are gradually converging with digital technologies, thanks to which, their champions proclaim, green technology will be ten times more effective. This raises the legitimate question of whether these technologies will in turn make green-tech pollution worse. This is certainly not what proponents of the energy transition extol. Digital technologies, they tell us, will in fact moderate our energy consumption. Such is their overriding message, which I will now carefully unpack:

- First, digital technologies form the basis of 'smart' electrical grids, which are designed to optimise our energy consumption. Between solar panels that generate clean energy and 'zero-emission' cars that use energy without polluting, there need to be grids to supply energy. With the standard grid, the electricity generated by coal-fired, oil-fired, and nuclear power plants feeds continuously into the grid. Because we determine output, we know almost exactly how much energy passes at any given time and at any point in the grid. There is nothing of the sort with the current energy transition, which draws on what is known as 'intermittent' energy sources. 'Intermittent' because no one has worked out how to control the sun and the wind. The power supplied to the grid by solar panels and wind turbines is sporadic. Grid operators are therefore faced with the task of directing the right amount of electricity to the right place at the right time. If there is not enough electricity, supply comes to a standstill. If there is too much electricity, the surplus is wasted. Energy operators are therefore delighted by the prospect of a new generation of electric grids that can constantly and dynamically manage supply in response to actual demand — thereby reducing energy wastage — using increasingly sophisticated algorithms.

- Digital technology supposedly mitigates the carbon impact of human activities, as lauded by the invigoratingly optimistic writings of new-technology advocates such as Jeremy Rifkin.[14] He postulates that the crossover of digital technologies and green energy will allow anyone to produce their own clean, inexpensive, and abundant electricity. The techno-prophet returned to the scene a few years later with an incredible idea: the new 'zero marginal cost society'.[15] By creating a new generation of 'collaborative commons' whereby everything is shared and exchanged online, internet technologies will pitch us into the age of access trumping ownership. No longer will we need to own anything; by simply surfing the web, we will have the freedom to share any product in exchange for money. Already we are seeing the beginnings of this cultural revolution in car transport (Lyft, Blablacar, Car2Go, GoGet, to name but a few). The consequences could be grave for the automotive industry. According to Rifkin, 80 per cent of car-sharing website users have already sold their car. Just think of the plummeting numbers of cars in this new 'Age of Access', and the commodity and carbon-emission savings that will go with it![16]

- In 2013, Eric Schmidt, the then chair of Google's board of directors, and Jared Cohen, a former advisor to Hillary Clinton in the US State Department and the self-appointed father of 'digital diplomacy', took the rationale a step further with the publication of their book *The New Digital Age*.[17] The global bestseller has helped open our eyes to the growing role of the virtual world. Courtesy of the internet, the two gurus explain, 'the vast majority of us will increasingly find ourselves living, working, and being governed in two worlds at once': the physical world and the virtual world. In the future, there will be more and more cyber-states, declaring more cyber-wars against virtual criminal networks that perpetrate increasingly powerful cyber-attacks.[18]

> Yet this is also a prophecy that promises the utopia of a dematerialised world. Already dematerialisation is synonymous with working from home, e-commerce, electronic documentation, digital data storage, and more. By limiting the physical transportation of information, and migrating from paper to digital, we can disavow our resource-guzzling civilisation and, while we're at it, slow down the deforestation of the Amazon and the Congo Basin.[19] In a nutshell, we are parachuting into a wiser, more moderate age.

But digital technology requires vast quantities of metals. Every year, the electronics industry consumes 320 tonnes of gold and 7,500 tonnes of silver; accounts for 22 per cent (514 tonnes) of global mercury consumption; and up to 2.5 per cent of lead consumption. The manufacture of laptops and mobile phones alone swallows up 19 per cent of the global production of rare metals such as palladium, and 23 per cent of cobalt. This excludes the other forty or so metals, on average, contained in mobile phones. (See Appendix 6 for the rare metals composition of a smartphone.) And yet 'the product in the hands of the consumer only makes up 2 per cent of the total waste generated over the course of the product's lifecycle', explain the authors of a French book that delves into the dark side of digital technology.[20] One example says it all: 'The manufacture of just one 2-gram chip produces 2 kilograms of waste' — a ratio between the end product and the resulting waste of 1 to 1,000.[21]

And this is only the manufacture of digital devices. Operating electric grids will, of course, generate additional digital activity

— and therefore additional pollution, the effects of which are becoming clearer. A documentary investigating the environmental impact of the internet traces the journey of a simple email: once it leaves the computer, it reaches the modem, travels down the building to a connection centre, transits from an individual cable to national and international exchanges, and then goes through a message host (usually in the US). In the data centres of Google, Microsoft, and Facebook, the email is then processed, stored, and sent to its recipient. All told, our email travels 15,000 kilometres at the speed of light.[22]

All this has an environmental cost. 'The ADEME calculated the electrical cost of our digital activities: an email with attachment uses as much electricity as a high-wattage energy-saving lightbulb … for one hour,' explains the documentary. Every hour, some ten billion emails are sent around the world. That's '50 gigawatts, or the equivalent output of 15 nuclear power stations for one hour'. One data centre alone uses as much energy as a city of 30,000 inhabitants to manage the flow of data and run its cooling systems.[23]

A US study estimated that the information and communication technology sector consumes as much as 10 per cent of the world's electricity, and produces 50 per cent more greenhouse gases than air transport annually.[24] According to a Greenpeace report, 'were the cloud a country, it would be the world's fifth-biggest consumer of electricity'.[25]

This is just the tip of the iceberg, for the energy and digital transition will require constellations of satellites — already promised by the heavyweights of Silicon Valley — to put the entire planet online.[26] It will take rockets to launch these satellites into space; an armada of computers to set them on the right orbit to emit on the correct frequencies and encrypt communications using sophisticated digital tools; legions of super calculators to analyse the deluge of data; and, to direct this data in real time, a planetary mesh of underwater cables, a maze of overhead and underground electricity networks, millions of computer terminals, countless data-storage centres, and billions of tablets, smartphones, and other connected devices with batteries that need to be recharged. Thus, the supposedly virtuous shift towards the age of dematerialisation is nothing more than an outright ruse, for there is no end to its physical impact.[27] Feeding this digital leviathan will require coal-fired, oil-fired, and nuclear power plants, windfarms, solar farms, and smart grids — all infrastructures that rely on rare metals.

Yet not a word about this is uttered by Jeremy Rifkin. So I reached out to the illustrious thinker to discuss the material reality of this supposed invisibility and the paradox of green energy. Repeatedly, I contacted the Foundation of Economic Trends, through which he offers himself as a speaker and consultant. I sent letters — via email — requesting clarification on this contradiction. I also suggested a brief meeting with Mr Rifkin during one of his trips to France, and while we were in the Washington suburb where his offices are located.

My questions were left unanswered. Perhaps on account of the fatal error underlying the energy and digital transition: they were not ground-truthed. It's all very well that green tech is thought up in the mind of a scientific researcher, finds real-world applications through the efforts of a persevering entrepreneur, benefits from attractive tax regimes and lenient regulations, and has the backing of plucky investors and benevolent business angels. But we forget that all green technology begins prosaically as a gash in the Earth's crust. This new demand on the planet replaces our dependence on oil with an addiction to rare metals. We are offsetting deprivation with excess — a bit like drug addicts weaning themselves off cocaine by sinking into heroin. Instead of addressing the challenge of humanity's impact on ecosystems, we are displacing it, and the rate at which we are subjugating today's environmental perils could very well have dire ecological consequences.

The broken promises of recycling

Can we moderate our consumption by recycling rare metals on a large scale to mitigate their impact and that of their extraction?

The idea is so appealing that the Japanese have already started to put it to work. It's an autumn afternoon in 2011, and I am in the suburb of Adachi, in the north of Tokyo. The tranquil atmosphere is suddenly disrupted by a commotion of blue-bin lorries. One of their drivers, Masaki Nakamura, is collecting electronic waste — e-waste — and piles up all manner of old video game consoles, mobile phones, and television screens in the back of his lorry. At the end of his round, Mr Nakamura takes his bounty to the dump of the Kaname Kogyo recycling company nearby. We find its CEO, Matsuura Yoshitaka, complete in a dark suit and tie, clambering over the mounds of home appliances that are being meticulously sorted by his teams. 'Nowadays people throw out appliances without much afterthought,' Yoshitaka tells me amid the crash of metal being moved from tip to tip. 'Yet they're full of rare metals!'

Clearly, globalisation has hurtled us into an unprecedented era that has made Western countries so prosperous that even our waste makes us rich — be it food, household, industrial, nuclear, or electronic. We have gone from a world not that long ago where our grandparents still battled with daily deprivations, to a civilisation that is at a loss about what to do with its staggering surplus. From scratching our heads at how to manage what we have to consume, we are faced with the conundrum of what to do with what we've already consumed.[28] Take scrap metal: every year, humans produce the equivalent of 4,500 Eiffel Towers of electronic waste, or six kilograms of waste per individual.[29] This figure that has ratcheted up an alarming 20 per cent over the last three years worldwide.[30]

Until now, industry has contented itself with recycling the big metals, and has done so successfully: over 50 per cent of gold, silver, aluminium, and copper has already been reprocessed worldwide.[31] But no one has taken a real interest in the smaller, hidden metals. This is where Japan has broken new ground by recognising the sheer quantity of rare-earth metals contained in the thousands of 'urban mines' (e-waste dumps) littering its territory.[32] For example, every one of Japan's 200 million used smartphones contain a few tenths of a gram of rare metals that can be isolated. That's 300,000 tonnes of rare-earth metals potentially lying dormant across the island country — enough to keep it self-sufficient for the next three decades.

This realisation is the driving force behind one of the most innovative circular economies for e-waste. (See Appendix 10 on the lifecycle of metals.) Every year, collection drives are held across Japan to return 650,000 tonnes of small electronic items to the consumption circuit. The campaign is so popular that it has earned the endorsement of Japan's favourite virtual celebrities, including Hatsune Miku. Against a background of manga *Keitaï* (mobile telephones), the scantily clad anime popstar squeaks in electronic tones about the goldmine that is the country's waste.

But the campaign is not enough, and Tokyo has also invested hundreds of millions of dollars in scientific research to find substitute metals[33] and to reduce the rare-earth–metal content of its magnets.[34]

Other Western countries have begun to follow suit. The US army, for instance, a formidable consumer of rare metals, has warehouses on the outskirts of Tucson, Arizona, filled to the rafters with out-of-service aircraft, all concealing tonnes of rare-earth metals that US army generals are unable to extract or reuse.[35] Worse still is the fact that as the world's most powerful army withdrew from Afghanistan, it is believed to have left behind $6 billion worth of military equipment crammed with magnets that its enemies could dispose of as they pleased.[36] Many in the United States have recognised just how big a challenge it is, and have suggested giving soldiers instructions on how to retrieve components containing rare-earth metals before demobilising.

For industry, it's another matter entirely, for the circular economy has completely upended traditional supply chains. In addition to having to know where their raw materials are sourced, manufacturers must now also be able to trace the users of their product. Companies such as Apple and H&M, which of course know where their rare-earth minerals and cotton bales are coming from, now have to trace the billions of iPhones and old jeans scattered across the globe. In other words, the roles of sender and recipient have been switched.[37]

Doing things in reverse for the same outcome is, for many, a Copernican revolution. But it is also a way to increase the portion of recycled metals in the supply chain. It could also give us a glimpse into the future of rare metals: a world in which the mining powerhouses are not the countries with the richest mineral deposits, but those with the most bountiful waste dumps; where treasure maps of the world's biggest scrap mountains will be drawn, and some debris will be ranked 'world class' in the same way as some of today's mineral deposits. Our waste dumps will be coveted goldmines.

Accordingly, Japan hardly mines any rare metals from its soils. Its pre-eminence in this circular economy could also turn it into a powerful exporter of those retrieved metals on which other countries depend. Thus, the geopolitics of recycling would be born — at least, this is what Japan believes. It's also not hard to imagine the technological advances that would make production more ecological, reducing the need to mine and to export old television sets to e-waste dumps in Ghana and Nigeria.

While this ambition looks good on paper, it is incredibly complex to implement. Unlike traditional metals such as iron, silver, and aluminium, rare metals are not used in their pure state in green technologies. Rather, the manufacturers in the energy and digital transition are increasingly partial to alloys, for the properties of several metals combined into composites are far more powerful than those of one metal on its own. For example, the combination of iron and carbon gives us steel, without which most skyscrapers would not be standing. The fuselage of the Airbus A380 is in part composed of GLARE (Glass Laminate Aluminium Reinforced Epoxy), a robust fibre–metal laminate with an aluminium alloy that lightens the aircraft. And the magnets contained in certain wind turbine and electric vehicle motors are a medley of iron, boron, and rare-earth metals that enhance performance.

Today, there is a profusion of new materials — translucent concrete, paper bricks, insulating gels, and reinforced wood — that transform the properties of the original material. These alloys are so promising that green technologies will increasingly become dependent on them. But, as the name suggests, alloys need to be 'dealloyed' to be recycled.

Numerous technologies for dealloying rare metals already exist. I learnt about one of them at Toru Okabe's lab at the University of Tokyo in 2011. Amid a tangle of cables, pipes, and thermometers, the researcher demonstrated his latest invention: a high-temperature autoclave that uses salt from the high-altitude salt flats of Bolivia. 'We use the salt to selectively leach the rare-earth metals from other metals,' he told me.

Clearly, recycling an alloy is anything but straightforward. Understanding it takes us back to our bread metaphor. To save the loaf of bread left over at the end of the day, our baker will try to separate its ingredients — an insanely complex, time-consuming and energy-intensive endeavour. The process is no less difficult for the rare metal magnets found in wind turbines, electric vehicles, and smartphones: manufacturers have to use time-consuming and costly techniques involving chemicals and electricity to separate rare-earth metals from other metals.

If alloying is like marriage, recycling is like divorce: it comes at a price. 'The technology I'm showing you has a lot of potential, but it's far from financially viable,' Toru Okabe admitted. Thus, the rare metals in Japan's waste dumps are hidden treasures that no economic model today can retrieve. It is the prohibitive cost of recovering rare metals — a cost that currently exceeds their value — that is holding industry back. The price of recycled metals could be competitive were it not for the fact that commodity prices have been structurally low since the end of 2014.[38]

For manufacturers, there is little point in recycling large quantities of rare metals. Why rummage through e-waste dumps when it is infinitely cheaper to go straight to the source? It is not surprising, therefore, that only eighteen of the sixty most used industrial metals have a recycling rate above 50 per cent.[39] An additional three metals have a recycling rate over 25 per cent,[40] and three more a rate of over 10 per cent.[41] The recycling rate of the remaining thirty-six metals is below 10 per cent.[42] For rare metals such as indium, germanium, tantalum, and gallium, as well as certain rare-earth metals, the rate is between 0 and 3 per cent.[43] (See Appendix 7 for a summary table of rare metals recycling rates.) It would be a stunning achievement for manufacturers to one day be able to recycle just 10 per cent of rare-earth metals, as is Japanese group Hitachi's ambition.[44] Yet the volumes of recycled metals will still not absorb demand. Even recycling nearly 100 per cent of lead has not been enough to stop its mining and extraction, because of perpetually growing demand.[45] It would seem that hell is well and truly paved with good intentions.

Return to sender

Despite this, manufacturers agree that rare metal recycling can be profitable by accumulating enough volume to create economies of scale. The problem is that in addition to being difficult to recycle, these metals don't stay in one place.

In the state of New Jersey, Newark Bay offers more than just views of the handsome towers of Manhattan. It hosts myriad US companies specialising in recycling e-waste — or, rather, exporting it. Their proximity to the nearby coastal ports is not lost on Lauren Roman, an activist from the US nonprofit organisation Basel Action Network. For years, she has been scouring New Jersey, photographing the tracking numbers on containers loaded with IT waste. With these numbers, she then tracks their journey around the world.

The overwhelming majority of recycling companies are obliged to process e-waste in the countries of its collection. This is the very thrust of the United Nations Basel Convention.[46] Adopted in 1989, the convention prohibits the movement of waste that is considered hazardous, on account of its potentially toxic heavy metal content, to countries with lower environmental standards.[47] To date, 186 countries and the European Union are parties to the convention, but a handful of countries — including the United States — have refused to ratify it. This emboldens US recycling companies to export unusable e-waste. After years of investigation, Lauren Roman confirms without a shadow of a doubt that 80 per cent of e-waste produced in the United States is sent to Asia.

This situation is not unique to the US. Japanese recycling companies also export their e-waste to China, despite being a party to the Basel Convention. One murky yet well-established line of business is *Kaitori*: waste buyback services that flout international regulations by loading used equipment into containers dubiously labelled 'humanitarian aid'.

Europe isn't doing any better. Countless 'second-hand' vehicles full of rare-earth metals leave the docks of Amsterdam. Adding to this loss is 50 per cent of used catalytic converters, a monumental stock of wind turbine batteries, over 50 per cent of used electronic cards, and a million tonnes of copper per year. Despite Europol classifying, from 2013, the illicit trade of waste as one of the biggest threats to the environment, nothing has been done about it:[48] European authorities estimate that up to 1.5 million tonnes of e-waste are exported from the European continent to the rest of the world, in particular Africa or Asia.[49]

Despite Beijing's ban in July 2017 on importing certain waste containing metals, China remains a major destination for electronic goods via Hong Kong. Chinese labour is so inexpensive that the cost of recycling is up to ten times cheaper than in developing countries.[50] As for magnets that cannot be recycled at an acceptable price, my Chinese contacts tell me that they are stored pending an economically viable solution — a claim I was unable to verify.

What can we conclude?

- 'Green' technologies require the use of rare minerals whose mining is anything but clean. Heavy metal discharges, acid rain, and contaminated water sources — it borders on being an environmental disaster.

Put simply, clean energy is a dirty affair. Yet we feign ignorance because we refuse to take stock of the end-to-end production cycle of wind turbines and solar panels. 'It's no longer enough to look at the green, sound, and non-polluting end product,' stresses Chinese environmental activist Ma Jun. 'We need to take a good look at whether the sourcing and industrial manufacture of the components making up the technology are environmentally friendly.'

- These very energies — deemed 'renewable' because they use energy sources (such as sunshine, wind, and tides) to which we have unlimited access — rely on mining resources that are not renewable. The wealth of minerals in the Earth's crust is finite, and the billions of years needed for their formation is no match for the exponential growth of our needs. I will revisit this point.

- In reality, these energies — still qualified as 'green' or 'carbon-free' because they help us wean ourselves off fossil fuels — depend on activities that produce greenhouse gases. It takes prodigious quantities of electricity from power plants to operate a mine, refine minerals, and transfer them to a manufacturer to be used in a wind turbine or solar panel. The result is tragically ironic: the pollution no longer produced in urban areas thanks to electric cars has merely been displaced to the areas where the resources needed for this very technology are mined. The energy and digital transition therefore belong to the wealthy. The most affluent areas are relieved of pollution by dumping its impacts on the most destitute and remote areas. But what can we do if we're not aware there is a problem in the first place? This is where our energy model is particularly pernicious: in contrast to the carbon economy, whose pollution is undeniable, the new green economy hides behind virtuous claims of responsibility for the sake of future generations

- The technologies lauded by many in ecological circles as the ticket out of nuclear depend on resources (rare-earth metals and tantalum) that release radiation when mined. While rare metals are not radioactive in themselves, the process of separating them from other radioactive minerals — such as thorium and uranium — to which they are naturally associated produces radiation in amounts that cannot be overlooked. Expert accounts put the ambient radioactivity of the toxic reservoir in Baotou and at the bottom of the Bayan Obo mining district at twice that recorded in Chernobyl today.[51] And despite the low level of radioactivity of normal mining conditions, International Atomic Energy Agency standards still require the generated waste to be isolated for several hundred years.[52]

- To accelerate the energy and digital transition, a few well-meaning souls would like to apply the concept of 'local loops' — used for the management of food products — to energy distribution. Eco-suburbs, such as Vauban in the German city of Freiburg im Breisgau, pride themselves on using only clean energy that is increasingly local. But has anyone thought to add the millions of kilometres travelled by the rare metals, without which these areas would not exist? 'Eco ideas like these work in theory. But in practice they are completely misguided,' a concerned expert tells me.

- Some of the green technologies underpinning our modest consumption ideals ultimately require more raw materials to be mined than do older technologies. A report by the World Bank states that 'a green technology future is materially intensive and, if not properly managed, could bely the efforts … of meeting climate and related Sustainable Development Goals'.[53] Denying this reality could very well lead us to go against the goals of the Paris accord. It could also result in a shortage of exploitable resources: the global population of 7.5 billion individuals over the next three decades will consume more metals than the 2,500 generations before us.

- Lastly, recycling the rare metals at the heart of our greener world is not as eco-friendly as we are led to believe. Its environmental footprint could even increase as alloys become more complex and involve more materials in greater proportions. Industrial companies in the energy and digital transition will have to contend

> with an immovable contradiction: their quest for a more sustainable world could effectively stifle the emergence of new, modest consumption models hinged on the principles of the circular economy. Perhaps future generations will look back and say: 'Our twenty-first century ancestors? They're the ones who took metals out of one hole and put them in another!'

These observations may seem obvious to those in the business of commodities. But to the vast majority of us, they are so counterintuitive that it may be many years before we fully apprehend and admit to them. Come that day, we will pick apart decades of myth and illusion. We will listen more intently to the warnings of the Cassandras. Like Carlos Tavares, the head of France's biggest car manufacturer, PSA, who, at the Frankfurt International Motor Show in September 2017, warned of the harmful effects of electromobility on the environment: 'If we are instructed to make electric vehicles, authorities and administrations must assume the scientific responsibility of this choice. Because I don't want to find that in 20 or 30 years, we've fallen short in aspects such as battery recycling, how we make use of the planet's scarce resources, or the electromagnetic emissions we generate when recharging batteries.'[54]

Perhaps we can expect an 'electricgate' that, like the 'dieselgate' scandal, will result in legal proceedings on a global scale. We will wonder how we could have turned a blind eye to the mounting evidence. We will admit that the consensus between economic and political circles, and backed by numerous environmental groups, drowned out any arguments to the contrary. We may even reach the conclusion that nuclear energy is in fact less harmful than the technologies we wanted to replace it with, and that we'd be hard pressed to do without it in our energy mix.

We will need to design new technologies, which some will undoubtedly qualify as 'miracle technologies', to rectify the immense problems this heedless march towards a greener world will have caused.

And we will not be able to lay the blame on the Chinese, Congolese, or Kazakhs alone. Westerners have brought this situation on themselves by knowingly allowing the most irresponsible countries to flood the rest of the world with dirty metals.

CHAPTER THREE
Delocalised pollution

RATHER THAN TAKE THE LEAD IN RARE METALS, THE WEST CHOSE TO shift its production and accompanying pollution to poor countries willing to sacrifice their environment for financial gain.

I decided to see this for myself. One morning in 2011, leaving the buzz of Las Vegas, I headed south-west along Interstate 15 — that straight asphalt strip cutting across the vast expanses of Nevada and California. Two hours later, I found myself looking down a quarry in the midst of a corroded industrial wasteland, a tired star-spangled banner ruffling overhead. Until 1990, US mining company Molycorp operated in this depression of rocks and shrubs that is the Mountain Pass mine. It is also the biggest rare-earth mine on the planet.

When the US dominated the rare-earth metals market

Contrary to popular belief, rare metals reserves are not concentrated in the world's most active mining countries, such as China,

Kazakhstan, Indonesia, or South Africa.[1] They are spread across the planet in varying degrees of concentration, making them both rare and not rare. The most strategic of these mineral reserves — rare earths — are found in a dozen or so countries.[2] A report by the French parliament states: '[B]efore 1965, extraction took place in South Africa, Brazil, and India; but total production was marginal: fewer than 10,000 tonnes per year.'[3] Between 1965 and 1985 there was a second phase, during which the United States was the global leader in rare-earth metal mining. The report continues: 'Although they did not have the monopoly, they were clearly dominant, with quantities as high as 50,000 tonnes per year.'[4] Mountain Pass was the mine that supplied these resources.

But the environmental damage caused by Molycorp's operations began to have a serious impact on the surrounding ecosystems. Clearly, the situation was an embarrassment for group management, for our request to visit the quarry was flatly refused.

So, in an act of desperation, I contacted a light aircraft rental company and informed Molycorp that since I wasn't allowed to walk through the facility, I would enter their airspace the following day at noon. Accompanying me that day was John Hadder, executive director of a very vocal environmental nonprofit organisation in the region, Great Basin Resource Watch (GBRW).

The next day, from the tarmac of a Las Vegas aerodrome surrounded by mauve-tinted mountains, our rickety plane was catapulted to altitude and, within minutes, we could make out the Molycorp mine in the middle of the mineral landscape. Soon afterwards we were flying over the rocks of the quarry and spiralling down to a body of water.

The most instructive part of the flight extended some 20 kilometres away from the excavation itself: a circular decant pond spread over several hundred metres in the heart of the desert. 'When the mine was in operation, all the discharged water was channelled to the pond,' John tells me. 'This contaminated water is still seeping into the ground water.'

Back on terra firma, in the shadow of a hangar next to the runway, the document John Haddar showed me was edifying: a map of Mountain Pass based on intelligence gathered between 1984 and 1998 by the United States Environmental Protection Agency. GBRW must have requested the declassification of this information to allow them to map the ecological damage caused by Mountain Pass's operations in the Mojave Desert.

What struck me was the succession of numbers reproduced on the map between the excavations and the decant pond. 'The water polluted by the ore processing was pumped out of the quarry and evacuated into this big pond,' Hadder explained. Directing the billions of litres of wastewater was a pipeline. 'The numbers shown here along the pipeline show where there were ruptures or leaks.' Over fifteen years, some sixty spills occurred. 'The worst was in 1992: a leak of one and a half million litres! In total, nearly four million litres of wastewater have spilled into the desert.'

This environmental damage hit local communities hard. The soil was contaminated with a noxious blend of uranium, manganese, strontium, cerium, barium, thallium, arsenic, and lead.[5] Polluted sand and contaminated groundwater fouled the Mojave Desert for miles around. 'After a series of lawsuits, the company was forced to tackle the environmental issues head-on,'

John told me. Molycorp was fined heftily.

Once, Molycorp was even paid a visit by armed federal agents, complete with shields and bulletproof vests, to notify the company of yet another violation of California's environmental protection regulations. And to protect the nearby habitats of the desert tortoise, the mine's 300 employees were ordered to attend training to learn about the reptile, and were prohibited from coming within 30 metres of them. At the end of the 1990s, fearing new spills and faced with the astronomical cost of modernising their facilities, Molycorp began to reassess the continuation of its operations at Mountain Pass.

Meanwhile, a China that was thirsty for economic growth saw in this challenging period for Western mining companies an opportunity to become a dominant player in the rare metals market. This was an ambition not without substance, for its mines in Baotou (see Chapter One) represented nearly 40 per cent of the world's rare metal reserves. To accelerate the shift of mining production from West to East, China used — and continues to use — formidable cunning that is captured in just one word: dumping. It engaged in trade dumping by slashing production costs; and environmental dumping because, as environmental activist Ma Jun explains, 'production costs do not factor in the cost of repairing the environmental damage'. And has China done anything at all to even paper over the damage?

Naturally, this dual-dumping strategy dragged down Beijing's prices, and by 2002 a kilogram of rare-earth metals produced in China had an average cost price of US$2.80 — half that in the US.[6] Molycorp did not stand a chance against this ruthless competition, and shut down its operations at Mountain Pass, running down its stocks until finally closing the mine in 2002.

An environmentally ethical approach would have been to set aside the quest for financial gain by subsidising, albeit at a loss, rare metals mining in Western countries where ecological responsibility held currency. This was a responsibility that could have fallen to Australia. In 2001, a wealthy Australian businessman, Nicholas Curtis, acquired Mount Weld — one of the biggest rare-earth deposits on the planet, located in the state of Western Australia. 'Curtis believed these materials would become highly strategic given the diversity of their applications, be it for cars, phosphors, or televisions. He founded the junior mining house Lynas to extract cerium, lanthanum, and neodymium from the Mount Weld mine,' said an expert on the matter.[7]

But the businessman had to face the fact that Lynas stood no chance of competing against the rock-bottom prices offered by the Chinese. The financial crisis in 2008 further jeopardised the opening of the mine.[8] And all the while, the West left the 'sorcerer's apprentice' of rare metals to his own devices. 'We in the West knew full well at what cost we were accessing rare-earth metals that were more or less clean, and which posed no risk to their future generations,' says a French expert. 'But we preferred to turn a blind eye to what was happening in China.'[9]

The Americans are not the only ones to have washed their hands of the affair. They share this responsibility with the French.

Hot-air balloons, adventure, and rare earths: the Rhône-Poulenc saga

France, a leading supplier of resources for the third industrial revolution? To jog our memory, let's take a look back at one of the shining periods of the humble television. In France, many will recall those Saturday nights in the 1980s at around 10.00 pm when everyone gathered around the box to watch *Ushuaïha*. The show's presenter was environmentalist Nicolas Hulot, France's answer to David Attenborough.[10] Viewers were transported to far-flung places to encounter little-known peoples, discover exotic animals that would have had Rudyard Kipling green with envy, and drift silently over epic landscapes in the helium-inflated envelope of a hot-air balloon. Broadcaster TF1 sold viewers the dream — while also selling off the little attention they had left. At the very bottom of the screen could be seen the rectangular logo of the program's official sponsor, French chemical company Rhône-Poulenc, together with the tagline: 'Welcome to the world of adventure, human feats, and exploits!'

Before its chemical arm became Rhodia in 1998 and it merged with Belgian group Solvay in 2011, Rhône-Poulenc was one of the two biggest rare metals chemical companies in the world.[11] In the 1980s, its factory in La Rochelle, in the west of France, purified between 8,000 and 10,000 tonnes of rare earths every year — that is, 50 per cent of their global supply.

This merits repeating: half of the most strategic rare metals — the resources of the future that would shape the energy and digital transition — were beneficiated in France. The country possessed unparalleled chemical knowledge coupled with superior commercial acumen. So much so, in fact, that French intelligence agencies monitored access at the plant just in case a Russian or Chinese guest on a working visit decided to indulge in a bit of industrial espionage.

The La Rochelle plant sits on a 40-hectare site along the coast west of the city. It is still in operation, and I was able to visit it in 2011. In the shadow of colossal warehouses, rare earths are separated and calcined in high-temperature furnaces to produce oxides (powders) that are packaged for the market. In the adjoining warehouses are row upon row of tightly packed bundles stamped with names we are now familiar with: cerium, dysprosium, neodymium, terbium, and so on. There is also an R&D laboratory, but the activity of the refinery itself is a shadow of its heyday twenty-five years ago.

It's no secret that extracting and refining rare earths produces high amounts of pollution. This is due to their natural association with radioactive elements, such as thorium and uranium. During separation at the La Rochelle plant, radon was released, albeit in minute quantities. Former employees of the French industrial company insist that inhaling the weakly radioactive gas has never had any impact on worker health. Yet '[t]here is no such thing as rare-earth minerals that are not radioactive'. And that's coming from a former director at Rhône-Poulenc.[12]

The few dozen tonnes of uranium that Rhône-Poulenc separated every year were sold to the national electricity utility EDF for its nuclear power stations. The thousands of tonnes of thorium were, and still are, stored at the plant in the hope that they could one day be used as fuel for a new generation of cleaner nuclear power stations. The effluent by-product of separating the minerals went to a wastewater treatment plant before being discharged into the sea via an outfall pipe located on the shore of Port Neuf, at the end of the bay of La Rochelle.[13] The liquid waste had a high concentration of iron, zirconium, aluminium, silicon, magnesium residues, and other impurities. In the 1980s, there were numerous incidents of untreated sludge 'escaping' from the wastewater treatment plant to be decanted directly into the sea.

Did these discharges contain radioactive thorium? One former Rhône-Poulenc employee says no: all thorium was stored upstream of the treatment plant — that is, before the evacuation of effluents. He added that any radioactivity would only have emanated from the radium found in thorium and uranium. We should also point out that Rhône-Poulenc went to great lengths to mitigate the inevitable pollution of its operations. But non-governmental organisations (NGOs) believe otherwise, and estimate that since 1947 the plant has evacuated some 10,000 tonnes of radioactive waste into the ocean.

In 1985, a prefectural order toughened environmental regulations and banned Rhône-Poulenc from discharging any effluents through pipelines not submerged at high tide, when waste could be swept away by the sea currents. The order also capped effluent discharges. Nevertheless, at the request of a local green political party, Les Verts Poitou-Charentes, the Commission for Independent Research and Information on Radioactivity (CRIIRAD) conducted several inspections of the facility from 1987. Contrary to official reports, they found that the outflow pipe was in service at both high and low tide. Samples taken in the immediate surroundings of the evacuation channel confirmed 'extensive discharges and an accumulation of thorium and its progeny in the marine environment', as stated in a letter from the nonprofit to La Rochelle's MP and mayor, Michel Crépeau. The radioactivity near the outfall pipe was at 1,000 counts per second, or 100 times the local average. 'It was a very serious affair,' says an engineer at CRIIRAD today.[14]

Not that this ruffled any feathers with the local authorities. 'Everyone knew [about the radioactivity], starting with the mayor Michel Crépeau. It was regularly reported on by the local press, but nothing was done about it,' says Hélène Crié, a reporter covering the events at the time for French daily newspaper *Libération*.[15] This contrasts with the position of Professor Pierre Pellerin, director of the French Radiation Protection Agency, which reports to the Ministry of Health: 'Every so often this business about radioactivity at La Rochelle crops up. If we keep up this nonsense, we are going to send everyone into a panic.'[16] Coming from the scientist accused of downplaying the fallout of the radioactive cloud that drifted over France after the Chernobyl disaster, one can't help but read this with circumspection.[17]

The environmental conscience of the French in the 1980s was a shadow of what it is today. But locals began to get hot under the collar. *Libération* wrote: 'Instructors at the local windsailing club worry that children sailing near that shore could accidentally swallow water or fall and hurt themselves in the contaminated mud. As for the bay as a whole, "because of the currents, it can take up to three weeks for a drop of water that's entered the harbour to be carried out."'[18] Committees were formed, and at public meetings community members could be heard chanting 'Rhône-Poulenc is an atomic timebomb!' and 'It's going to blow!'[19]

Member of Parliament Jean-Yves Le Déaut describes how he went to La Rochelle on two occasions to assess the situation — and was met by nearly 300 protesters with loudhailers. 'One gentleman came up to me and said: "We've been living a peaceful existence here, Mr Le Déaut. And now we are being exposed to radioactivity … People here are starting to get scared.'"[20] Jean-Paul Tognet, a former industrial and raw materials director of Rhône-Poulenc and Rhodia Rare Earths, recalls the 'growing criticism of Rhône-Poulenc's reputation by the media. Management wanted to pull the plug — the controversy almost shut down the La Rochelle plant.'[21]

From 1986 to 1998, Rhône-Poulenc was under the management of Jean-René Fourtou. In 1994, he changed tack entirely: 'I don't want to hear anything more about radioactivity. Buy whatever you need, but I will not allow a single radioactive product.'[22]

And with that, Pandora's box was opened. Rhône-Poulenc (which became Rhodia) found itself looking for foreign partners to carry out first-stage refinement. This is how, one fine day, the group asked Norway, India, and China whether they could make La Rochelle's non-radioactive products for them.

Were there errors in Rhône-Poulenc's mineral processing, and were the subsequent plant tours organised for the public too little, too late to redeem their lack of transparency? Did the residents of La Rochelle, with their limited understanding of radioactivity, exaggerate the risks to which they were actually exposed? And did the authorities raise suspicions by attempting to cover up certain information? No doubt everyone has their share of the blame. Either way, during this time other countries got ready to fill the void. 'In the early 1990s, the Norwegians supplied us with raw materials at high prices,' says Jean-Paul Tognet. 'We should have kept the diversity of our supplies, but instead we stopped working with the Norwegians and entered into a long-term partnership with the Chinese, who were more competitive.'

Naturally, buying cheaper improved the French chemical company's bottom line. It was also profitable for its customers, eager to procure transformed rare earths at rock-bottom prices. It made perfect sense. Jean-Yves Dumousseau, who was sales director at the US chemical company Cytec at the time, explains: 'Obtaining rare earths from anywhere else would have been far more expensive than continuing to procure supplies from China at a quarter of the price! It's the same argument for electronics, jeans … everything! I'm sorry to say it, but it was that simple.'[23]

Meanwhile, thousands of kilometres away from La Rochelle, China and its annual production of 100,000 tonnes of rare earths claimed the monopoly. As for working conditions, '[t]here were no checks on the separation units or even safety procedures; you'd find the guys in the refineries doing electrolysis at 700 degrees without hardhats, wearing flip-flops and shorts! It was outrageous!' says Dumousseau.[24] Another source put it more bluntly: 'It was a mess, and no one wanted to deal with it.'

A new world order

The Rhône-Poulenc saga brings to mind those who despair of where the world is headed. To find a sense of order in this 'chaos', they latch on to clean divides: countries of the north and countries of the south; countries of the east and countries of the west (once separated by the Berlin Wall); emerging countries and developed countries; the Orient and the Occident; the free world versus the axis of evil; 'Old Europe' against the vibrant 'New World'; and so on. But there may be an even deeper divide that began thirty years ago — one that tells the story of the world of our making. It did not follow a treaty in Versailles, a congress in Vienna, or a conference in Yalta. Rather, it is an industrial order that formed organically between China and the West (to polarise once more).

It was first formally acknowledged in the infamous 'Summers Memo', an internal document at the World Bank signed in 1991 by its chief economist, Lawrence Summers. He purportedly recommended that developed economies export their polluting industries to poor countries, and especially to 'underpopulated countries in Africa [that are] vastly under-polluted', as 'the economic logic ... is impeccable'.[25] Anxious to explain himself after the memo was leaked, Summers naturally pleaded intentional sarcasm. Yet his comments are perfectly aligned with the reality: as our societies strive towards 'zero risk', all kinds of industrial activities have steadily been pushed out of Europe and the rest of the Western world.[26]

Consider the example of REACH — the European regulation aimed at minimising the sanitary risks associated with over 30,000 chemicals found in consumer goods.[27] It has vastly improved the quality of life of the European Union's 445 million citizens, and especially that of industrial workers. Then, in 2007, the United States, Canada, and Mexico signed an agreement in Montebello, which also had a public health objective. Under the agreement, the signatories are required to list all chemical products on the North American market.[28] But these environmental safety requirements have also taken the wind out of the sails of Western industry, which has had to suspend the production of myriad chemical substances that are now banned.[29] Now, others are free to take over production. And become Europe's suppliers.

The same logic applies to green technologies. In the last two decades of the twentieth century, the workload of the future energy and digital transition fell naturally between China, which did the dirty work of manufacturing green-tech components, and the West, which could then buy the pristine product while flaunting its sound ecological practices. Thus, the world was ordered as Larry Summers intended: between the dirty and those who pretend to be clean.

This is precisely the takeaway of the Mountain Pass and La Rochelle sagas: by moving the sourcing of its rare metals to China, the West chose to relocate its pollution.[30] We have knowingly and patiently created a system that allows us to move our 'filth' as far away as possible, and the Chinese — far from pinching their noses — have welcomed the initiative with open arms. As magnanimously put by a Canadian rare metals industrialist: 'We can thank them for the environmental damage they have endured to produce these metals in our place.'[31]

At this point, we should steer clear of any anti-capitalistic arguments: the countries on our radar have adopted this system of their own volition by deliberately shaping their economies to generate massively inflated profits. Many have done well on the progress afforded by the globalisation of the markets. Except now the Chinese have realised something else. A Chinese academic explains: 'We are praised for moulding to the Western-ruled world order of the time. But China also suffered. And in terms of cost-effectiveness, I am not wholly convinced of the advantages in our favour.'[32] Under this arrangement, Beijing has effectively laundered dirty minerals. Concealing the dubious origins of metals in China has given green and digital technologies the shining reputation they enjoy. This could very well be the most stunning greenwashing operation in history.

Naturally, businesses in the West are complicit. 'They couldn't care less about the conditions under which the minerals are extracted and refined,' a European industrialist tells me. 'All that matters is having the lowest price possible.' They are also complicit in our ignorance of the human dramas at play behind the scenes of the energy and digital transition. In the 2018 annual report by US behemoth Apple, a major consumer of rare metals, the words 'rare earths', 'minerals', or 'metals' do not appear.[33] As for Tesla, the biggest name in electric vehicle manufacturing, its 2019 environmental report is extremely discreet when it comes to these resources. The artisanal cobalt mines in the Democratic Republic of Congo are mentioned, but nothing is said of their environmental impact.[34]

Consumers could have led the resistance: their choice to buy or boycott has the power to redirect the market and change its practices. They had the information: numerous documentaries have exposed the distressing environmental and social impact of electronic goods, as have countless NGO reports.[35] Consumers would have needed to put pressure on manufacturers to design more ecological products, such as the repairable Fairphone,[36] and to use their votes to pressure their governments to beef up feeble anti-obsolescence regulations. But consumers would hear nothing of it: a connected planet was better than a clean one. But there is hope. New consumer rights regulations have emerged, such as the law passed in 2015 in France that requires manufacturers to inform consumers about the availability of spare parts. In the United States, the 'right to repair' movement is gathering pace; it requires electronic goods manufacturers to allow consumers to fix their products by providing them with repair manuals, for instance.[37]

Such citizen-led initiatives to repair domestic appliances have mushroomed. Another one is the Restart Project. Since its creation in 2013, the London-based social enterprise has organised thousands of events around the world where participants learn how to repair their electronic goods.[38] And in the US, iFixit is an online platform with over 50,000 free tutorials on how to repair anything from mobile phones to connected cars.[39]

Today, Europe and the US want their whites whiter than white, with the EU setting out the ambitious 2030 climate and energy framework,[40] and the US beating its carbon dioxide emissions-reduction record in 2017.[41] Yet how could Europe possibly hope to achieve its targets if its polluting industries came home? Environmental responsibility has already been transferred, and it's not a transaction we will be reversing anytime soon.[42] Therefore, is the West — brandishing its self-declared legitimacy — truly in a position to lead talks on the fight against global warming? Shouldn't COP 21 have been held in Beijing, Kinshasa, or Astana rather than in Paris? And is France really in a position to urge, as President Macron did, that we 'Make Our Planet Great Again'?

The truth is that we are no better than companies that boast colossal earnings to their shareholders — while hiding a mountain of debt in an obscure subsidiary in the Caribbean. These 'off-balance-sheet' transactions that use fraudulent accounting practices have led to the conviction of countless company directors. We glorify our modern ecological legislation — while shipping out our electronic scrap to toxic waste dumps in Ghana, exporting our radioactive waste to the ends of Siberia, and outsourcing rare metals mining around the world, to make a deadweight loss look like a net profit.

The illusion of a new era of opulence

The zeitgeist of the 1990s played an important role in this reshuffling of roles. At the start of the decade, George Bush and Margaret Thatcher gave wing to the expression 'peace dividend'. The fall in military spending at the end of the Cold War gave rise to a new era of peace and economic prosperity. One can recall the feeling of optimism — euphoria even — that reigned at the time. With the end of the nuclear arms race came demilitarisation, and the states that had built up stockpiles of rare metals in preparation for an armed conflict were left wondering whether they were worth holding on to.[43]

A strategic stockpile is like a savings account: when the forecast looks gloomy, we tuck some money away for that rainy day. Naturally, in the bright days of optimism, we dip into our savings and enjoy the sunshine. So, in the 1990s, we witnessed a global sell-off of strategic stockpiles.[44] In France, the platinum and palladium stockpiles deposited in the safes of Banque de France were quickly sold off by both left- and right-wing governments. The dozens of billions of dollars' worth of lithium and beryllium rare earths in the US went the same way.[45]

This cornucopia largely originated in the former USSR and its satellite states. Palladium stockpiles were sold discreetly and in record numbers via Zurich. As a former asset manager relates: 'It started with the banks UBS and Credit Suisse, working for end-consumers in the jewellery sector, for instance. Palladium was also bought by trade majors like Glencore and Trafigura.'[46] Meanwhile, Beijing pursued a quite different strategy of building its reserves while buying up a handsome portion of the market inventory.

The sudden superabundance of raw materials on offer sent prices into a long-lasting decline that gave the impression of limitless availability — an illusion of an era of abundance given substance by the breaking down of international commodity trade barriers. In the mining sector, businesses cared only about buying metals as cheaply as possible.

It was only a matter of time before this elusive opulence diverted industrial companies from their duty of knowing the origin of the resources they relied on and managing their supply. This is precisely what happened in the lumber industry. In Europe, many producers and artisans no longer know the exact origins of their materials. As production lines have scattered and the distances travelled by timber have grown longer, parquet floor layers, carpenters, and the like have become disconnected from the substance of their trade. The problem is that the day China suddenly decides to snap up French oak, for example, no one will know where to find new suppliers.[47]

It's the same story for perfumes. From the 1950s, globalisation and the low cost of labour led to perfume-makers neglecting their flower farms in Grasse, in the south of France, in favour of less noble products from Egypt, India, or Bulgaria. 'Perfume was sold based on concepts rather than on the quality of ingredients,' a professional tells me.[48] It's a philosophy that the perfume sector is now reconsidering.

This is also more or less what happened in the industries that use rare metals. The assumption still exists that resources are available in any quantity and at any time. Supporting this utopia are the dogmas of 'just in time' and 'zero stock'. Taught in MBA programs the world over, and applied by big industrial groups, the two production management methods were established in 1962 by Taiichi ōno, an engineer at the Japanese company Toyota (hence the term 'toyotism'). 'Just in time' is about avoiding surplus stock by making the time between the manufacture of a product and its sale as short as possible, resulting in what is known as 'lean manufacturing'. Its corollary, 'zero stock', outsources the management of spare parts and components to a multitude of contractors — thereby transferring the risks associated with the delivery of raw materials to these third parties.

This complexification of supply chain logistics means that the exact origins of a raw material are not wholly transparent from one end of the supply chain to the other. Instead, participants can see no further than one level upstream and one level downstream. Surely this explains why, in a recent annual report, under the heading 'raw materials risk', Thales — a multinational whose rare metals–intensive electronics are used in aerospace, defence, security, and land transport — sanctimoniously states: 'Given the nature of its business, Thales uses few raw materials. The Group's exposure to raw materials risk is therefore negligible.'[49]

Hypocrisy? Ignorance? Either way, toyotism has helped relieve businesses of their responsibility with respect to 'metals risks'.

An end to public policies on mineral sovereignty

We see this same insouciance and lack of foresight within governments, which have steadily put mining strategies on the backburner. However, before the collapse of the communist bloc, this was not the case for France, as illustrated by the French Geological Survey (BRGM) — a public institution world-famous for its mining expertise. The BRGM even benefited from the oil crises of 1973 and 1979, which also served to familiarise French political leaders with the realities of resource scarcity. In 1978, as an initiative of the French minister of industry, André Giraud, the government launched the 'Metals Plan' — a vast mining stocktake aimed at boosting the BRGM's activities. Former employees nostalgically refer to this period as the 'golden age of BRGM', when it had a mining exploration program that enjoyed the active support of the government, especially in metropolitan France and French Guyana.[50] Its exploration department had 250 employees, and offered its services in francophone Africa, Portugal, and Quebec.[51]

In the 1990s, the Metals Plan came to an end, and the dynamic BRGM began to lose its lustre. By 2000, exploration activities were wrapped up, marking the start of what would be called 'the winter of French mining'.[52] 'To think we had gold, zinc, tungsten, antimony, and silver,' recalls a former employee, 'but investors were becoming increasingly rare.'[53] The mines that hadn't already been shut down were abandoned, creating their own lot of social dramas. 'The mining industry never made up a large part of French GDP,' says another former employee, 'but when you add up the few hundred employees per mine, you're left with a lot of people. At the end of the day, we lost an entire industry. Not to mention our mineral sovereignty, our capacity to supply our industries with our own minerals and metals.'

This underlying trend is common to all Western countries. One need only look at infographics on the history of global mining production. While Europe produced nearly 60 per cent of the world's heavy metals in 1850, its momentum steadily declined to produce no more than 3 per cent today. Mining production in the US hasn't fared any better: after peaking in the 1930s, accounting for close to 40 per cent of global production, it now represents around 5 per cent.[54]

Back in Europe, the three French White Papers on defence and national security produced in 1971, 1994, and 2008 made no reference to the supply of rare metals, despite how critical these materials are to military technology. Only in the 2013 publication did the term first appear.[55] It was of little interest to French intelligence agencies. 'The government never asked the agencies to take any action in this regard. I think the Directorate-General for External Security was light years away from such matters,' admits former intelligence director Alain Juillet. According to several corroborative sources, however, in the 1970s a former spymaster, Alain de Marolles, convinced the head of French intelligence, Alexandre de Marenches, to add eight strategic minerals to a metals supply risk list. De Marenches subsequently stepped down, de Marolles was 'honourably discharged', and their cause went no further.[56]

Three decades were enough to bring about an about-turn in strategy. Up until then, a nation's power hinged largely on its ability to rely on its own vital resources — or, failing that, to do everything in its power to guarantee supply from outside its borders.

Consider the example of oil. In the UK at the start of the twentieth century, the First Lord of the Admiralty, Winston Churchill, made the decision to convert the Royal Navy from coal to oil. He also did this to secure oil supplies from the Middle East for his country. The British government acquired a controlling interest in the Anglo-Persian Oil Company, and crisscrossed Persia with gigantic marine pipelines.

As for the United States, when they realised after the Second World War that their own oil reserves would not be enough to meet their growing energy needs, they turned to the Kingdom of Saudi Arabia and its extraordinary crude oil reserves. The 'Quincy Pact', signed on 14 February 1945 between President Roosevelt and the Saudi king, Ibn Saud, gave Washington privileged access to Riyad's petroleum in exchange for military protection. France looked to Algeria and Gabon in the same way. When it came to food, however, Paris always managed to safeguard part of its sovereignty during World Trade Organization (WTO) talks by limiting the liberalisation of agricultural markets. Just as it safeguarded its civil nuclear program to maintain its energy sovereignty …

For millennia, the fundamental rules of relying on one's own resources or securing a sustainable supply beyond one's borders have dictated every strategy for achieving energy autonomy. Yet to date, neither of these has been applied to rare metals. One could argue that the quantities at stake are inconsequential when compared to our gluttonous appetite for oil. But as we have by now discovered, these metals are as discreet as they are indispensable. Even though every person on the planet consumes as little as 20 grams of rare-earth metals a year, the world would run infinitely slower without them. And yet few futurists have looked into how important these minute metals have become as a result of our technological choices since the 1970s. The policy of both demanding and claiming complete dependence on others is now widely accepted, whereas not that long ago it would have been considered utterly ruinous.

This was certainly before the market became so short-sighted, as this US expert says: 'Western countries no longer have long-term strategies, and rare metals are no exception.'[57] There is also the particular French context. France's mines and its abundant agricultural and fishing resources have made it less averse to dependence than other countries, like Japan, that have had no choice but to develop a trading culture and to find reliable supply routes to compensate for their lack of natural resources. And since France is not a country of traders either, it has not developed a culture of economic intelligence as far as the rare metals market is concerned. Says one specialist: 'The French DNA is not equipped for a situation of scarcity.'[58]

In short, the Western world honours the 'cargo cult' founded not too long ago in the Pacific Islands. Between the end of the nineteenth century and the 1940s, the mosaic of Melanesian peoples scattered predominantly across Papua New Guinea, the Fiji Islands, and New Caledonia came into sudden contact with Western societies — first with French and British colonisers seeking gain and conquest, and then with the US Army during the Pacific War. Both sets of fresh occupants of the region needed regular supplies of food and non-food items, so they built logistics networks that linked the spray of islands to the rest of the world.

Just imagine the stupefaction of those ancestral peoples upon seeing the arrival of boats and then planes, their holds loaded with treasures. And their amazement at how easily these goods appeared: all it took were radio operators to fire off their requests for medication, rations, and equipment to land on the fine-sand beaches or be parachuted down from the sky, as if by magic. Naturally, the Melanesians did not have the slightest notion of the industrial fabric woven behind these supplies. But because one apparently needed only to ask in order to receive, they began to imitate the Westerners, designing dummy radio devices, laying fake runways, watching, and waiting — for a very, very long time — for their needs to be met. The Westerners gave these rituals the name 'cargo cult'.[59]

In the twenty-first century, on the other side of the world, our societies — as informed and materialistic as they are — are giving in to a similar cult. The genius of logistics has stripped us of a fear that obsessed our ancestors for 70,000 years: that of scarcity. But everything comes at a cost: the globalisation of supply chains gives us consumer goods while taking away knowledge of their origin. We have gained in buying power what we have lost in buying knowledge. Is it any wonder that 16 million adults in the United States think that chocolate milk comes from brown cows?[60]

But not everyone is displeased as the West sleepwalks its way forward. For by organising the transfer of rare metals production, we have done much more than palm off the rare metals burden to the galley slaves of globalisation. We have entrusted a precious monopoly to potential rivals.

CHAPTER FOUR
The West under embargo

CELEBRATIONS WERE IN FULL SWING. THE WEST CONGRATULATED itself on its new ecological conversion, while deep in Jiangxi province the Chinese slaved in the gullies to crush the ore needed for this very transition. The lowliest work went to China, while in the West we focused on high-value-adding industries. We were winning at the game whose rules we ourselves had written and imposed.

That was before some geologists armed to the teeth with quantified reports turned up to break up our party. Indeed, this endangered species — for their numbers have dwindled in step with mining activities in the West — confronted us with an unpleasant and inconvenient reality. China, a preponderant producer of certain rare metals, could for the first time ever decide to cut off exports to the countries that most needed them.

Beijing: the new rare metals master

Every year, the United States Geological Survey, a government agency of the United States Department of the Interior, does an assessment of mineral resources for the preparation of its vital *Mineral Commodity Summaries.* In the report, analysts pore over ninety raw materials considered critical to our modern economies. Its 200 pages provide detailed statistics on resource availability, global stocks, and, most importantly, where in the world resources are mined.

The latter statistic is alarming: the 2017 report finds that Beijing produces 44 per cent of the indium consumed worldwide, 55 per cent of the vanadium, nearly 65 per cent of the fluorspar and natural graphite, 71 per cent of the germanium, and 77 per cent of the antimony.[1] This concurs with the findings of the European Commission: China produces 61 per cent of the silicon and 67 per cent of the germanium, 84 per cent of the tungsten, and 95 per cent of rare-earth metals. The conclusion from Brussels is sobering: 'China is the most influential country in terms of global supply of the majority of critical raw materials.'[2]

Following China's lead, myriad countries applying a specialist mining strategy have also acquired majority, if not monopolistic, positions. The Democratic Republic of Congo produces 64 per cent of the world's cobalt; South Africa 83 per cent of the world's platinum, iridium, and ruthenium; and Brazil 90 per cent of the world's niobium. For its beryllium needs, Europe is dependent on the United States, which accounts for over 90 per cent of production. There are also countries whose share of global production is substantial enough to trigger temporary shortages and wild price swings. For instance,

Russia alone controls 46 per cent of the world's palladium supply, and Turkey 38 per cent of the world's borate supply.

Having the upper hand on rare metals is a question of survival for Beijing; the United States is not the only country very concerned about its supply security.[3] The reason is that China is not only the world's biggest minerals producer, it is also the world's biggest minerals consumer.[4] To meet its needs, it guzzles as much as 45 per cent of global industrial metals production[5] — an appetite equally voracious for agricultural commodities,[6] oil, powdered milk, and even Bordeaux wine. (See Appendix 9 for China's share of global consumption of certain commodities.)

Chinese strategists are well versed in the challenges of mineral sovereignty. As a student in France, Deng Xiaoping worked in an iron foundry of Le Creusot.[7] As for his successors: 'The last six presidents and prime ministers — apart from the current prime minister [Li Keqiang], who read law — were all trained in engineering — electrical, hydroelectrical, geology — and in process chemistry,' a natural resources strategist tells me.[8] Wen Jiabao — Hu Jintao's prime minister from 2003 to 2013 — was a geologist by training. Consequently, and with the support of a stable authoritarian political system that values patient and consistent decision-making, Deng Xiaoping and his successors were able to lay the foundations of an ambitious policy to secure the nation's supplies.

The method was comparable to that of a steamroller: in the space of a few decades, China multiplied the number of mines in its territory, and launched works for a second 'silk road' on land and sea as a raw materials supply route from Africa, while acquiring and merging companies in the commodities sector. Global markets and the geopolitical status quo were steadily upended as Beijing's sphere of influence grew. More than simply taking its place on the global rare metals market, China underwent a complete transformation to become a maker of these markets.

It has become so powerful that any decision made in Beijing today has unavoidable global repercussions. One dip in local mining production, and the well-oiled wheel of supply and demand jams. A sudden jump in domestic demand is enough to trigger massive shortages. This is what happened with titanium — a mineral 50 per cent supplied by China globally. An unexpected rise in Chinese consumption between 2006 and 2008 increased titanium prices tenfold,[9] and put French aircraft manufacturer Dassault Aviation in a serious supply predicament.[10]

Chinese foreign policy's 'weapon of choice'

Beijing quickly realised the power it possessed from its stranglehold on rare metals. To put this into perspective, we can look at OPEC, the Organization of the Petroleum Exporting Countries. For decades, its fourteen members have been able to significantly influence the barrel price, yet they represent 'only' 41 per cent of global oil production. China, on the other hand, has staked its claim on 95 per cent of global production of the coveted class of certain rare-earth metals. In the words of an Australian expert: 'It's OPEC on steroids.'[11] So what does a nation do when it realises just how powerful it is? Naturally, its intentions begin to take on a far more aggressive hue.

This is precisely what China is doing. The precepts of a hostile rare metals trade policy were reportedly outlined by Deng Xiaoping during a tour of the Bayan Obo rare-earths mine in the spring of 1992. 'The Middle East has oil; China has rare-earth metals,' he presaged. Chinese businessmen are known to smugly quote these words at raw-materials meetings and summits — words that say all that needs to be said.

At the beginning of the 2000s, close observers of the rare metals market noticed something not quite right. China's export quotas, set at 65,000 tonnes in 2005, began to drop a year later to just under 62,000 tonnes. By 2009, Beijing had further reduced this to 50,000 tonnes, and official figures for 2010 put exports at only 30,000 tonnes.[12] The same trend was observed for all the rare metals disproportionally produced by China. For instance, in August 2001, China enforced quotas on its molybdenum exports to the European Union. It then imposed a series of exorbitant export taxes between 2007 and 2008.[13] The WTO's analysis of the complaints against the Chinese was unequivocal: over the previous two decades, China had engineered a policy of systematic restrictions on rare-mineral exports, ranging from fluorspar, coke, bauxite, magnesium, manganese, yellow phosphorous, silicon carbide, and zinc.[14]

In the 2000s, everyone from Jakarta, Los Angeles, Johannesburg to Stockholm began to feel the pinch from China. 'Every month, we worried about what quotas we could next expect,' says Jean-Yves Dumousseau, who worked in China at the time.[15] The hardest hit by the Chinese offensive was Japan — a major consumer of rare earths on account of its high-tech industries. Speaking anonymously, a Japanese diplomat I met in Tokyo told me: 'I attended many meetings between the Japanese minister of industry and Chinese government representatives. Whenever we raised the issue of rare earths, the Chinese made it plain that they could at any point turn off the tap [of exports].'

'There was no doubt that sooner or later a much bigger crisis would blow up in our faces,' an expert in the industry confirms. After all, isn't the twentieth century full of examples of states with a dominant position in strategic resources imposing embargos for commercial, diplomatic, or military gain?

One such example takes us back to the 1930s, when the United States imposed a helium embargo (of which it was the sole producer) on Germany for fear that the Nazis, already using the gas for their Zeppelin airships, would eventually put it towards aggressive ends.

Then, in 1973, OPEC declared an oil embargo against Israel and its allies in response to the Yom Kippur war, sparking the first oil crisis in history.

In 1979, US president Jimmy Carter halted the export of 17 million tonnes of grain to the Soviet Union in response to its invasion of Afghanistan.[16]

And, more recently, Russia cut off its gas exports to Poland and Ukraine due to diplomatic tensions — a dispute covered extensively by the international press.[17]

After gas, oil and then grain were wielded as weapons, it was inevitable that China would weaponise its metals. And so, in September 2010, it launched an embargo on rare earths that defies belief.

It was also to be the first embargo of the energy and digital transition.

Trade manoeuvres with global repercussions

Behind the massive trade fallout is a longstanding dispute between Japan and China over the Senkaku (or Diaoyu) Islands — an archipelago of five small islands and three rocks in the East China Sea, north-east of Taiwan. While it might not sound like much, the area conceals vast amounts of oil and gas, which is why the two Asian powers have coveted the islands since the end of the nineteenth century.

Japan took the Senkaku Islands from China after the First Sino-Japanese War in 1895. They were then placed under US control at the end of the Second World War, before being handed back to Japan in 1972. This has not stopped China from asserting its sovereignty over the territory — and incurring the wrath of the Japanese. So, when a Chinese trawler dared to cast its nets near the coasts of the islands on 7 September 2010, the Japanese coastguard saw this as a provocation and gave chase. The scene that followed — filmed and available online — is captivating: refusing to comply, the Chinese captain steered his trawler into a collision with the Japanese patrol boat.[18] His detention by the coastguard caused an uproar in China. Quick to tug at any nationalistic thread, the Chinese media had no difficulty in sustaining the nation's indignation over the incident.

Was it any surprise, therefore, that two weeks later, on 22 September, all deliveries of Chinese rare-earth metals to Japan were halted, and without any official declaration of an embargo? 'The incident with the trawler sharpened our nationalistic instincts,' Chen Zhanheng, vice secretary-general of the China Rare Earth Industry, told me in Beijing. 'Several Chinese companies took it upon themselves to suspend their deliveries to Japan!' Astonishingly,

Chen Zhanheng made no bones about the fact that the Chinese authorities, not wanting to upset the WTO, denied that any official embargo had been imposed.

A year after the events, Japanese businesses still did not believe a word of this. So I headed 2,000 kilometres away from the Forbidden City to Japan. Pulling out of Tokyo station in the Shinkansen high-speed train, I slowly skirted round Mount Fuji and its conical outline jutting into the autumn sky. Four hours later, I began to make out the tentacles of Osaka, the country's third-biggest city, extending out along the Pacific coast. I had come to talk to Kunihiro Fujujita, a rare-earth metals importer, at his factory warehouse to hear his version of events.

'China has always used its natural resources as political leverage,' he said. Wearing a dark suit and a hard hat, Fujujita showed me his stockpile of yttrium — a rare earth used for precision electronics. But in September 2010, his orders were suddenly no longer honoured by his Chinese suppliers. 'The Japanese industry was in a panic,' he said. Rare-earth metals are the 'vitamins' of its high-tech industry. They are so vital that 'even a cleaning lady knows what they're about'.[19] What had started as a banal incident at sea turned into a catastrophe for Japan.

Soon the crisis took on an international dimension. In the days that followed, many European and US rare metals importers also began to worry about the sharp decline in Chinese exports. The Western media, up till then largely unfamiliar with these tiny metals, sunk their teeth in as well, headlining 'international tensions', a 'showdown' between China and Japan, and a 'war' over the acquisition of this 'crucial category of minerals'[20] that are 'more precious than gold'[21] and used in cutting-edge industries. The EU trade commissioner's spokesperson stressed that rare earths were a 'major concern' for the European Commission, and appealed to China to 'allow the markets to operate without hindrance'.[22] The US secretary of state, Hillary Clinton, spoke on the matter at a press conference in Hawaii, and announced an imminent visit to China to resolve the crisis.[23] Some weeks later, Jean-Louis Borloo, the French environment minister under the Nicolas Sarkozy presidency, published a ministerial order to create the Committee on Strategic Metals to address the risks of shortages of metals critical to French industry. And on the steps of the White House, President Obama announced America's submission of a trade complaint to the WTO against China.

The rare metals war had begun.

Back in Osaka, Kunihiro Fujujita tells me how he continued to suffer the effects of the 'informal' embargo for six months after the arrest of the captain of the trawler. That is, until it occurred to China that it, too, might find itself short of high-tech consumer goods 'made in Japan' that could no longer be exported due to a lack of resources![24] But in the meantime, the rare metals market was gripped by panic as it was hit with the reality of the supply shortage. This, and the hand-wringing over Beijing's moves, together with the speculative behaviour of certain Chinese traders, caused prices to skyrocket, with a host of other rare metals following suit.[25] Brokers, traders, and importers in all four corners of the globe spent most of their time trying to extract any semblance of a promise to deliver the materials or, failing that, passing on the consequences of these unprecedented disruptions to unsuspecting customers.[26]

'It was utter chaos!' says Fujujita. 'The natural course of supply and demand no longer had its place.'

Journey to the queen of platinum

The new 'metals risk' is not linked to China's export policy alone. A surge of nationalism over mining resources is sweeping across Asia, Africa, and Latin America, and is increasingly weakening Western positions.

Nowhere is the groundswell more apparent than in some of the most hidden parts of southern Africa. Johannesburg, the economic capital of South Africa, was my next port of call. From there, I drove three hours along vast expanses of savannah to Phokeng, in the North West Province. At first glance, everything in this isolated town of 22,200 inhabitants in the middle of the bushveld appeared normal. Yet it was anything but. Gigantic totems and multicoloured flags bearing the official emblem of the crocodile reached into the sky, while a group of men wearing the blue uniforms of the Royal Bafokeng Reaction Police Force patrolled the pristine streets.

There was no customs house or border post. Nor any sign indicating that, just a few kilometres earlier, I had crossed the invisible border of the Royal Bafokeng Kingdom — a territory covering 1,400 square kilometres, almost the land area of Greater London.

While the kingdom is an integral part of the 'rainbow nation', it has also established its own governance, administration, clan structure, budget, and customary law system. The reason for this, and also why I had come, lies several hundred metres underground:

the most extraordinary deposits of platinum group metals — ruthenium, rhodium, iridium, platinum, and more. These rare and precious metals have multiple applications, ranging from jewellery and laboratory equipment to catalysts for cars.

From Phokeng, we cut a few kilometres north through the veld to reach the Rasimone mines, its ground pocked and swollen by the extraction of rocks from below the surface. Standing before us, in the middle of the bush against a rolling landscape, were the towering iron behemoths of the platinum refineries. Railway tracks interlaced the facilities, their carts of waste rock hauled by trundling locomotives overhead.

'When I started here, we only mined on one level. Now we're ten levels down!' exclaimed Dirk Swanepoel, a white South African miner working for Anglo Platinum. A stone's throw from his office, a dozen or so miners wearing overalls and hard hats climbed off a conveyer belt coming out of a chasm made in the rock. They had been using pneumatic rock drills to bore holes that would subsequently be blown up using explosives. The ore brought to the surface is crushed and reduced to particles of rock containing platinoids. They are then plunged into water containing special reagents, decanted, dried, melted, and purified to obtain platinum. 'Every month, we extract 200,000 tonnes of rock,' Swanepoel told us. 'Each tonne contains about four to seven grams of platinum.'

The Bafokengs happen to be sitting on the biggest platinum deposit in the world — a treasure that the 'People of the Dew' were unaware of when they settled in the fertile lands in the fifteenth century. In 1870, King Kgosi Mokgatle acquired the first 900 hectares of the current territory, using the fortune amassed by his subjects in the diamond mines of neighbouring Kimberley. The precious metals were discovered in 1924.

After the birth of democratic South Africa, the Bafokengs — wronged by the segregationist regulations of apartheid, which prohibited them from owning their land — entered into a long legal battle with Impala Platinum (Implats), which had been pocketing the proceeds from mining on the territory. The Bafokengs emerged triumphant, and now receive 22 per cent in royalties, and even took a stake of nearly 13 per cent in the company.[27] It was the first time a South African ethnic community had come up trumps against a mining company.

Leruo Molotlegi, who has been the *kgosi* (king) since 2000, and the thirty-sixth monarch of the dynasty, is overseen by the queen mother, who has a symbolic role. Semane Molotlegi is one of the last queens of the 'dark continent'. She is virtually invisible in the media, and it is almost unheard of for a foreign journalist to meet her. I was fortunate to be granted an audience with her.

I dressed up to the nines to meet with the exquisite woman in her fifties, who was draped in elegant, colourful traditional dress, her voice the essence of poise. She is known respectfully as *Mmemogolo* (meaning 'grandmother' in Setswana). We spoke about her travels around the world to spotlight the success of her people.

The Bafokeng contradict the curse of raw materials, by which Western societies ship in, use up the local resources, and ship out. Instead, the Bafokeng became the richest tribe on the continent, and even plan to diversify their sources of income.[28] The strategy they implemented is textbook-worthy: by standing up against mining companies, the people of platinum asserted the supremacy of the producer over the buyer, and of the owner — sovereign over its resources — over customers around the world. This case is also unprecedented in that it was a mere 'tribe', rather than a state, that took on a multinational.

While the Bafokeng baulk at the idea of resource nationalism, this is nevertheless a case study of a rebalancing, if not a reversal, of the traditional power relationship. The mining companies, often acting on behalf of Western consumers and used to imposing their terms, understood they would increasingly be outplayed and would therefore need to adapt. The landmark example of the Bafokeng captured the attention of international organisations, and the nation was paid a visit by the World Bank, the World Economic Forum, United Nations agencies, and US academics.[29]

The resurgence of mining nationalism

This is far from being an isolated case. Today, more and more states are refusing access by foreign industrial companies to promising mining areas. In 2013, the Mongolian government halted the operations of Rio Tinto in the Oyu Tolgoï copper mine in the Gobi Desert.[30] Other countries banned foreign companies from buying local mining companies: in 2010, the Saskatchewan province in Canada stymied an attempt by Anglo-Australian company BHP Billiton to buy out the Canadian group PotashCorp, the world's leading potash producer. States are also investing in traditionally private mining groups. This is what Qatar did in multiplying its holdings in mining groups such as the Swiss multinational Xstrata via the state-owned Qatar Mining Company.[31] The minute Gulf state seems to be on a quest for metals for which it has no apparent need.[32]

Lastly, and most importantly, restrictions on the free trade of metals are on the rise. Indonesia, like China, offers another eye-opening example. In 2009, the undeniable mining powerhouse declared a series of embargos on the export of more or less all the raw minerals produced by the archipelago. In 2014, it upped the ante by including nickel, tin, bauxite, chromium, gold, and silver. 'We had to ensure the sovereignty of our raw materials,' explains a senior government official I met in Jakarta. 'Any political action concerning our mining wealth must be determined and executed by our government and not by foreign states.'[33]

There are many more examples, as observed by the Organisation for Economic Co-operation and Development (OECD). Its most recent report on trade in raw materials gives an inventory of all basic product export restrictions declared around the world, and identifies 900 such cases between 2009 and 2012.[34] The report includes a fascinating chart that shows how measures have increased since 1961: they were at relatively low levels until 2005, but then the curve went up sharply, and has not come down since.

We see these restrictions in almost all the minerals and metals included in the OECD inventory. Argentina has imposed barriers on the export of thirty-seven mineral resources, while South Africa has done the same with copper, molybdenum, platinoids, and diamonds; India with chromium, manganese, iron, and steel; Kazakhstan with aluminium; and Russia with tungsten, bauxite, copper, and tin. What is driving this widespread trend?

Most mineral-producing countries have developed to become

emerging countries. Governments now have to consider their burgeoning middle class and the more prosperous and resource-hungry consumers they represent. The growing sentiment is that locally mined resources should be used for domestic consumption rather than to satisfy the appetites of buyer countries. There is also a growing ecological conscience and activism — once the domain of the West — that oppose local mining plans. This has resulted in states having to tighten social and environmental regulations, which makes it take longer to bring a mine into operation.

More broadly speaking, a culture of resistance is taking root from Jakarta to Ulaanbaatar, from Buenos Aires to Pretoria, as a newer and savvier middle class is wiser to what they see as a sell-off of their resources. They are spearheaded by political leaders who observe an economic rebalancing between developed countries, often bogged down in stagnation, and dynamic emerging countries craving wealth. Protectionist measures are therefore demonstrations of power in a 'de-Westernising' world.

This is not new: from the 1960s, the wave of independence in the third world came with claims for sovereignty over resources.[35] In Africa, in 1958, Ghana nationalised the Ashanti goldmines that were formerly under British control. The Democratic Republic of Congo followed suit in 1965, as did Tanzania in 1967, Zambia in 1970, and Zimbabwe in the 1980s. Then came the liberal wave of open trade. But it took the Chinese policy of slapping quotas on rare metals exports to reignite — and amplify — resource sovereignty across five continents. 'China galvanised the nationalism of resources,' says an American expert, 'not only on its own territory, but all over the world.'[36] From that point, it was no longer a question of *if* new trade crises would occur, but rather *when* they would occur.

Between taking sips of tea in the bar of a luxury hotel in the Shanghai suburb of Xuhui, Vivian Wu, a highly authoritative voice in the rare metals industry, tells me that such a scenario has every chance of coming true: 'Rather than embargos, I prefer to talk about action measures against Japan and other countries. These actions form part of a strategy led by the Chinese government to restore our image. And they may be applied again in the future, be it for rare earths or other metals.'[37]

Metals of influence, metals of crisis

The specific qualities of the rare metals markets could make matters worse:

- As we have seen, they are incredibly narrow markets: output is derisory compared to that of the markets for major metals such as iron, copper, aluminium, and lead. For instance, the global production of rare earths barely makes up 0.01 per cent of steel production.[38]
- They are highly confidential, and involve very few buyers and sellers. Any move on their part therefore is all the more likely to upset the balance of supply and demand. Accordingly, a default by just one supplier can throw consumers into a tailspin, just as a new technology requiring rare metals can cause a sudden shortage in supplies.

- They are opaque markets in which business discretion and an absence of formalities are the rule. Other than a handful of metals listed on the London Metal Exchange, there are no official reference prices, and everything is traded over the counter. Buyers often need to consult specialised journals or connect to Weibo, the Chinese microblogging website, where brokers and traders drip-feed information about their latest transaction amounts.

- Moreover, these markets are strategic for mining countries. China has proven very reluctant to provide certain production data that are considered state secrets.[39] Hidden stockpiles and geostrategic and diplomatic factors render the market even more obscure, even for leading analysts.

- And, finally, the sudden increase in private investors, in it for their own interests, has also impeded the interplay of supply and demand. For all resources, a specialist explains, these third-party players trade 'sixty times more "raw material" stocks than ten years ago', which has contributed to price instability.[40] Traditionally, the major base metals were the subject of market speculation, but increasingly the rare metals market is, too.[41] Speculators include 'hedge funds [such as the Tudor Fund in the US], asset managers [such as PGGM Investments in the Netherlands], pension funds [such as Pacific Investment Management Company (PIMCO) in the US], as well as the finance

> departments of US universities [such as Harvard and Princeton]', according to a banking analyst. Any major position taken in a metal results in wild speculation. One of many examples:[42] the purchase in 2017 of 17 per cent of global cobalt production — that is, several thousand tonnes — by investors betting on a shortage of cobalt brought on by a sharp price increase.[43]

Making any kind of forecast about these ultra-sensitive markets is therefore near impossible. 'The rare-earths market is neither stable nor predictable,' Vivian Wu emphasises. Gone are the days when, even without regular supplies of rare metals, governments and businesses could at least count on constant market prices to sketch out a strategy. As a specialist from a European geological institution says: 'Rare metals are crisis metals.'

CHAPTER FIVE
High-tech hold-up

MONOPOLISING MINING WAS BEIJING'S FIRST VICTORY. BUT IT WASN'T long before China began to turn its sights to downstream high-tech industries that use rare earths.

Battle of the super magnets

This process started with magnets, the metals of which I first encountered in July 2011 when visiting Peter Dent, the director of magnet manufacturer Electron Energy Corporation, at his company's headquarters in the small town of Landisville, Pennsylvania.

During my visit, Peter guided me past the warehouses to a storage area bathed in pallid neon lighting. 'There you have it: rare earths!' he exclaimed as he turned on the timer light. Dotted across the polished concrete floor were greenish barrels containing lumps of grey and slightly corroded material: samarium samples, gadolinium pellets, and other metals with perfectly unpronounceable names. After years of tracking these metals without ever laying my eyes on a single one, I felt like a buccaneer before Blackbeard's bounty. I had every intention of taking my time to explore this treasure trove.

More interesting still were the adjoining workshops. Dent continued the tour: 'This is the machine room.' In its din, workers spent all day machining small, round electromagnets containing rare metals. 'This is where we shape and size our magnets. We produce hundreds of thousands of them every year,' Dent said. After a well-oiled manufacturing process, the magnets were lined up on trays like bread loaves fresh out of the oven, and carefully loaded into carts.[1]

Until the mid-1970s, there were only a handful of industrial applications for rare earths and other rare metals — for their luminescent properties, they were used in lighters and camping lanterns.[2] Their scope of application first grew with the advent of colour television screens.[3] The real game-changer, however, was rare-earth magnets: developed in 1983, these pure marvels of technology have become indispensable in all products equipped with electric motors, reputed to be pollution-free.[4]

We know that an electrical charge coming into contact with the magnetic field of a magnet generates a force that creates movement. Traditional magnets made out of the iron derivative ferrite needed to be massive to generate a magnetic field powerful enough for more sophisticated applications. 'Remember how we used to walk around with mobile phones the size of bricks?' jokes an expert. Just one of the many shortcomings of an oversized magnet.

Meanwhile, in the mobility industry, the race for lighter and more efficient energy had begun. Engines needed to be as light and compact as possible — reducing the calibre of an engine meant reducing the proportions and weight of the object in which it was used. Such progress would without a doubt generate massive energy savings.[5]

This very progress was made possible by rare-earth magnets, revolutionising modern electronics. In fact, without even realising it, you have probably already come into contact with these super magnets, especially if you happen to have a magnetic knife-holder on your kitchen wall. Have you ever wondered how a single magnet can defy gravity to suspend a 20-centimetre steel blade in the air? Certainly not with the help of ferrite. 'At equal magnetic strength, a rare-earth magnet is 100 times smaller than a ferrite magnet,' an industry expert explained. 'This is about miniaturisation; rare earths make objects smaller.'[6] It's also about making electric engines powerful enough to challenge the dominance of internal-combustion engines. It gave the energy transition and digitalisation a formidable kickstart.

This is also precisely where the trouble started.

Rewind to the 1980s. Rare-earth magnets are all the rage, and have colonised manufacturing sectors the world over — giving Japan, and its electronics company Hitachi, which holds the patent, the unassailable lead in the industry. So much so, in fact, that 'the Japanese banned the export of this technology to China', Chen Zhanheng told me.[7]

Beijing wasn't put off in the slightest by the embargo. It decided that, in addition to making off with virtually all rare-earth resources, it would start to take control of the miracle technology behind the end products. In this way, explained Chen Zhanheng, 'China's own industries could benefit from the added value of rare-earth minerals.' And by whatever means.

In the 1980s, magnet manufacturers were mostly situated in Japan, from where they supplied the bulk of global demand. But they could no longer resist the siren song of their Chinese counterparts, who offered to take over the menial work of machining the least-sophisticated magnets. 'The Chinese said, "Come to Canton! Relocate your low-value rare-earth applications, and we'll take care of your low-tech!"' an Australian consultant told me.[8]

The Japanese may have had the technology, but the Chinese had the allure of cheaper production that would allow Japanese businesses to expand their profit margins. The Japanese didn't hesitate for long. Boasting full employment and a strong currency at the time, the island nation considered it a sound decision. The history books will look back at this and say that Japan, the second-strongest global power at the time, knowingly exported to its competitor the technologies it lacked.

As related in Chapter Three, French chemical company Rhône-Poulenc was enticed by the prospect of low-cost rare-earths transformation, and also moved part of its refining operations to China. It did this by setting up joint ventures with Chinese partners from the 1990s. This had trade unions up in arms, and it would have taken an insurrection and a fight to the bitter end to keep their jobs in France. But Rhône-Poulenc had other priorities. Its pharmaceutical branch was about to be privatised to become Aventis. Therefore, 'the strategic and geopolitical importance of this small [chemical] business was completely lost on them', recalled a former employee.[9]

Jean-Paul Tognet, who worked at Rhône-Poulenc at the time

and made several trips to China at the end of the 1970s to look into future partnerships, knew what was brewing: 'The technical assistance our partners wanted from the West was a one-way street: they expected everything to be handed over to them. They found it completely normal to be helped … without giving anything in return.'[10] He assured me that Rhône-Poulenc did not divulge any industrial secrets. But by abandoning their refining sector to the Chinese, only to then become their most loyal customers, the West — and France in particular — handed China the market on a silver platter.

In the 1990s, low-end refineries began to mushroom — first in the Baotou region and then in the rest of China. 'Across the country, rare earths had become the goose that laid the golden egg. Money flowed, and factory bosses drove around in Lincolns!' Jean-Paul Tognet recounted.[11] Essentially, we had created the ecosystem our opponents needed to reproduce Western knowledge, make massive profits, invest in their own R&D activities, and thus break into the downstream industry at full force. Jean-Yves Dumousseau, a chemist who worked in China at the time, shot straight from the hip: 'Rhône-Poulenc slipped China's foot into the stirrup.'[12]

But where was the harm? After all, Rhône-Poulenc believed it had a twenty-year head start on China. By discontinuing the more trivial aspects of its refining business, it could move up the value chain to more sophisticated intermediary products (particularly in luminescence). But in 1987, one of its engineers noticed the staggering progress being made by Chinese refiners, and put his foot in it: 'I told them we were only three years ahead of the Chinese at best. It caused an uproar!'[13] Only his timeline was off.

'By 2001, Chinese refiners were all on a technological par with us,' Dumousseau said.[14] 'We might have underestimated this competitive risk for too long,' Tognet ventured. 'The Chinese wanted to move up the value chain, and we couldn't stop them! We cared more about rock-bottom production costs, especially with customers breathing down our necks ... And of course, today nothing has changed.'[15]

Today, only a few rare earths are still transformed in La Rochelle. The separation workshops have closed their doors, and the site's operations are a shambles. The group's revenues have decreased, and headcount, which was at 630 employees in 1985, has halved.[16] Can Rhône-Poulenc (today Solvay) at least take pride from new downstream prospects that sustain jobs in France? 'Even those businesses were transferred to our facility in China ... It all comes down to cost!' Tognet said. Perhaps we can take comfort in the fact that Solvay still has interests in joint ventures with its Chinese partners. This may be, but, as Tognet bitterly put it, 'Solvay now sees itself as a Chinese company as well!'[17]

China's awakening set in motion unavoidable economic disruptions. But did we really need to give Beijing that extra leg-up? Our poor assessment of China's competitive streak, combined with our outright quest for profit, without a doubt precipitated the transfer of labour, work units, and, most importantly, technologies to China.

A chronicle of impending de-industrialisation

Underlying our apparent blindness are numerous instances of wishful thinking. The West has long held to the illusion of eternal scientific progress — a philosophy that has irrigated several economic sectors since the 1980s. We believed that by abandoning our heavy industries, we could focus our efforts on high-value-adding manufacturing sectors while maintaining healthy profit margins. Some believed that emerging countries would continue to be the factories of the world, churning out jeans and toys, while we would hold sway over the more lucrative segments. 'I think most people continued to believe that only blue-collar jobs would be affected by the shifts [brought on by Chinese competition],' a trade unionist from the US metal industry told me. 'We didn't realise that we were going to lose more than coffee-cup production, and that the more skilled jobs would be hit much harder economically.'[18]

Added to this is the pipedream of manufacturing fading into the background in favour of a service economy. The focus should be on knowledge and the immense added value it generates. This belief, echoing the utopia of a dematerialised world discussed in Chapter Two, was heavily subscribed to by the business world at the turn of the twenty-first century. Like Alcatel-Lucent's CEO Serge Tchuruk, many US and European business leaders could not resist the allure of 'factoryless companies'. Grey matter was more valued and therefore given more support, to the detriment of the tool that is the lowly factory. This line of thinking led to the separation of business from factory, as the former opted for outsourcing. In France, this trend was largely driven by the love lost between citizens and industry. 'At the start of my career,' Régis Poisson, a former engineer at Rhône-Poulenc, told me, 'a factory worker could become famous for something he designed. Then came the rejection of enterprise and the deterioration of the company image. Today, the working class no longer likes factories, which symbolise exclusion.'[19]

Thus, the West and China took this path hand in hand. But from the 2000s, the Chinese began to use far less conventional methods: rare metals quotas that rattled the magnet manufacturers that hadn't relocated their factories (to guard their industrial secrets). As their rare-earths provisions ran dry, they were faced with a difficult choice with equally painful outcomes: keep their industrial activities at home at the risk of having to slow down production for lack of raw materials, or relocate to China and have access to abundant supplies.[20] For Japan, the dilemma was short-lived, a London analyst explained to me: 'They were starved of raw materials. So they took their technologies to China.'[21]

Beijing's treatment of those determined to hold their ground was cruel: price distortion. This was much to the fury of Sherrod Brown, a US senator from Ohio, who, in a fiery speech in 2011, said, 'China is organising artificial shortages and export quotas to raise prices internationally while keeping them low domestically. How can we compete when the Chinese are so brazenly cheating?'[22]

This tactic was equally unjust for the vast majority of magnet manufacturers outside China. At the end of the 1990s, Japan, the US, and Europe made up 90 per cent of the magnet market, but now China controls three-quarters of global output. Using blackmail in the form of a technology-versus-resources trade-off, China expanded its monopoly in mineral production to include mineral transformation. It therefore dominated not one but *two* stages of the industrial value chain. This is confirmed by my Chinese expert Vivian Wu: 'I'd even go as far as saying that in the near future China will have an entirely integrated rare-earths industry, from one end of the value chain to the other.'

In fact, this wish has already partially come true. And nowhere more so than in the city of Baotou, in Inner Mongolia.

Journey to the 'Silicon Valley of rare earths'

In Chapter One we saw behind the scenes of the world's rare-earths capital, Baotou. We saw its toxic lakes and its cancer villages, whose inhabitants are dying a slow death. Let us now examine its glittering exterior.

It's a Saturday, and by day the city's elegant glass towers cut a sleek figure against the mineral desert. By night, when darkness falls over the plain, 'the little Dubai of the steppes' adorns itself in blinking lights and glowing lanterns, repelling the cold obscurity of the surrounding landscape. Jianshe Road, Baotou's main stretch, fills up with people who have come to admire the shop window displays and to peruse the fragrant food stalls lining the pedestrian alleys. There is a sense of triumph and conquest here; I can see it in the smug smiles of passers-by, and even in the buildings still wrapped in plastic sheeting — all symbols of a victorious metropolis convinced of its fantastic destiny.

It's an idyllic setting for a delegation of eighty international businessmen invited to Baotou to attend an international conference on rare earths in October 2011. Their hosts are clearly out to impress, having put up their guests in a luxury hotel with suites overlooking the verdant parks of the city centre. Even the cloudless skies seemed to have been ordered for the occasion.

In the conference room, Sun Yong Ge, a high-ranking official in charge of the Baotou Economic Development Zone, continues the charm offensive: 'Baotou is the rare-earths capital! We welcome technology industries — we can supply them with virtually all the minerals they need.'

The city has made technology the cornerstone of its development, its centrifugal force being its close proximity to the rare-earth deposits where China takes its fill. 'We want to be more than just suppliers of raw materials; we want to supply more elaborate products,' Yong Ge says. Western businesses that, like the colonisers before them, sought only to mine resources to generate added value back home are no longer welcome in Baotou. However, the apparatchik says, 'we are very open to transformation companies wanting to move their technologies to China'.

Whether out of interest or for lack of choice, numerous foreign manufacturers have already converged on the mineral refineries in the 120-kilometre free zone on the city's outskirts. An international presence is reflected in the numbers: according to Yong Ge, ever year Baotou produces 300,000 tonnes of rare-earth magnets — one-third of global production. As it happens, the delegation was invited to visit the plants. Journalists, however, were kindly requested to stay behind. So I entrusted a small camera to Jean-Yves Dumousseau, who came back with images of a power display. 'This is where they make the magnets you have in your iPhones and iPads,' he commented as we watched the video. 'The Chinese paid an absolute fortune for these technologies copied from European know-how.'[23]

By orchestrating the transfer of magnet factories, the Chinese accelerated the migration of the entire downstream industry —

the businesses that use magnets — to the Baotou free zone. 'Now they've moved onto producing electric cars, phosphors, and wind turbine components. The entire value chain has moved!' confirmed our witness. Sun Yong Ge added to the list: '10,000 tonnes of polishing material, 1,000 tonnes of catalytic converters, and 300 tonnes of luminescent materials.'

This makes Baotou much more than just another mining area. The Chinese prefer to call it the 'Silicon Valley of rare earths'. The city hosts over 3,000 companies, fifty of which are backed by foreign capital, manufacture high-end equipment, and employ hundreds of thousands of workers who generate revenues of up to €4.5 billion every year.[24] At this rate, boasted Sun Yong Ge, 'in ten or so years, our standard of living will be similar to that of the French'.[25]

When China decided to assume the burden of the 'oil of the twenty-first century' three decades ago, it did not focus on manufacturers on the decline nor turn its back on the high-tech sector. It chose to cross swords with the West over resources so that, one generation later, it could set its sights on the high-end digital and green-tech industries. Thus, rare-metal restrictions did more than serve China's sporadic embargos. The second stage of its offensive is far more ambitious: China is erecting a completely independent and integrated industry, starting with the foul mines in which begrimed labourers toil, to state-of-the-art factories employing high-flying engineers. And it's perfectly legitimate. After all, the Chinese policy of moving up the value chain is not dissimilar from the viticulture strategy of winemakers in the Napa Valley in California, or the Barossa Valley in South Australia. As one Australian expert put it, 'The French don't sell grapes, do they?

They sell wine. The Chinese feel like rare earths are to them what vineyards are to the French.'[26]

This strategy of moving up the ladder is not limited to rare earths. As early as the 1990s, a wave of concern rippled through the fabric of German small-to-medium-sized enterprises (known as the Mittelstand) specialising in the manufacture of machine tools. (Machine tools are used for factory work automation, from basic milling machines to ultra-connected machining centres.) The Mittelstand acted pre-emptively by gradually replacing humans with robots, allowing it to remain competitive to the extent that German industry still represents 30 per cent of the country's GDP.[27]

Except that industrial robots require terrific amounts of tungsten. China has always produced this rare metal in abundance, but there are other tungsten mines around the world, ensuring supply diversity for manufacturers. During the 1990s, the Chinese machined their own cutting tools — 'Some hammers, a few drills … really crumby tools,' said an Australian consultant.[28] But they wanted to move up the value chain in this area as well. 'They drove down tungsten prices [from 1985 to 2004],[29] hoping that Westerners concerned about getting their raw materials at the best price would buy exclusively from the Chinese, and that competing mines would shut down.'[30]

We can guess what could have happened next: the Middle Kingdom — now the hegemonic power in tungsten production — would have used the same blackmail tactic to force the Germans to move their factories as close as possible to the raw materials. The Chinese would have crushed any German lead in the cutting-tools industry, and would then have made off with the machine-tools segment — a pillar of the Mittelstand. It would have been the hold-up of the century!

But the Germans saw the Chinese coming, and aligned instead with other tungsten producers (Russia, Austria, and Portugal, among others). 'They preferred paying more for their resources to sustain the alternative mines and not depend on the Chinese,' the Australian consultant told me.[31]

No matter. China applied the same modus operandi to the graphite market, where it is also a notoriously preponderant player. The mineral graphite, the purest form of carbon, is used to produce graphene —a nanomaterial a million times finer than a single strand of hair, yet twice as strong as steel. Its discovery by physicists Andre Geim and Kostya Novoselov earned them the Nobel Prize in Physics in 2010.[32] Recognising the massive markets that graphite was creating, China 'pursued a strategy similar to that of developing the downstream value chain', explained Vivian Wu. China's commercial policy already includes export duties and quotas to protect its domestic market.[33] Recognising the danger, the US filed a new complaint at the WTO in 2016 against Beijing's practices that 'disadvantage US producers by raising the prices of these raw materials for manufacturers outside of China, while lowering the prices paid by China's manufacturers that use these same raw materials'.[34]

By now a pattern is emerging, and it is being applied to molybdenum and germanium, a journalist I met in Beijing told me.[35] Lithium and cobalt should go the same way.[36] 'They're using the same industrial policy for iron, aluminium, cement, and even petrochemical products,' warned a German industrialist.[37] In China, there is even talk of applying this policy to composite materials — new materials resulting from alloys of several rare minerals. What if it developed a miracle composite material that the rest of the world could not do without? China certainly wouldn't sell it any less sparingly than its other resources. In addition to keeping a blacklist of critical minerals, the European Union should list all critical alloys whose supply could be threatened.[38]

The West is starting to put words to what has happened with China: whoever has the minerals owns the industry.[39] Our reliance on China — previously limited to raw materials — now includes the technologies of the energy and digital transition that rely on these raw materials. 'Is this a non-military conflict? The answer is most assuredly yes!' said one US rare-earths expert.[40] Are we on the winning or losing side? The answer of a specialist from the French mining industry is cutting: 'We're not even putting up a fight!'[41]

Unsurprisingly, other mining states around the world are following China's lead. A case in point lies in the middle of the island of Java, in the Indonesian capital of Jakarta.

Indonesia's new 'non-aligned movement'

If there is one capital in the world that could be given the title of Hell City in the twenty-first century, it would Jakarta. Located in the immense archipelago of Indonesia in South-East Asia, and nicknamed 'The Big Durian' in reference to the foul-smelling fruit eaten by Indonesians, Jakarta is a city you endure rather than visit. And of all the senses vexed by this intolerable megalopolis of 30 million people, touch was without a doubt the most affected: the muggy heat saturating the air; the unrelenting rainfall harassing this city of glass and concrete; and the close shaves of my motorcycle taxi with the hordes of vehicles tearing down the main roads.

Because it was impossible to distinguish north from south, or to familiarise myself with any particular landmark — be it a tower, intersection, or major road — I mentally choreographed my own: cable-constricted bamboo trees; an armada of motorcycles straddling the torrents of water gushing out from the drains; the pungent smell of garlic invading the palm tree–lined alleys; the cadence of a train trundling overhead; the vestige of a forest forgotten between two buildings. And on it went.

A few days later, while coming in to land at Bangka, 400 kilometres further north, I had my first glimpse of the purpose of my trip that winter in 2014. From the plane, I saw thousands of craters — as if a meteor shower had rained down on this island the size of the Greater Paris area. They are in fact tin mines, below whose surface thousands of miners play their part in a thriving black market. There are also mines offshore, recognisable by the thousands of small wooden barges on the surface. From these crude vessels, young men plunge 20 metres deep with nothing more than a breathing tube connected to an air compressor. They scrape the seabed, sending the sand to the surface, with the help of a makeshift vacuuming device. From their barges, raw material is separated from minerals using a small sorter.

Bangka is the world's biggest producer of tin — a grey-silver metal essential to green technology and modern electronics, such as solar panels, electric batteries, mobile phones, and digital screens.[42] Every year, over 300,000 tonnes of tin are mined around the world. Indonesia represents 34 per cent of global production, making it the biggest exporter of this high-tech mineral, which is nevertheless not considered rare. The archipelago recognised the value of this outstanding mineral: from 2003, as a spokesperson for one of Indonesia's biggest mining houses, PT Timah, explained: 'Tin became the first mineral to be used in an embargo.'[43]

It would be the first of a very long series of embargos. From 2014, all of Indonesia's mineral resources — from sand to nickel, and diamonds to gold — were no longer exported in raw form. As explained by Indonesian authorities, 'The minerals we don't sell now will be sold tomorrow as finished products.' As in China, this policy was a powerful way to generate wealth. By some calculations, preserving the added value in this way quadrupled profits on iron, increased profits on tin and copper sevenfold, bauxite profits by a factor of as much as eighteen, and nickel profits by as much as twenty.

The Indonesians did far more than replicate the Chinese model: they innovated by laying the foundations of financial nationalism. In 2013, Jakarta established the Indonesia Commodity and Derivatives Exchange (ICDX) to fix the tin price, shunning the 'diktat' of the world's biggest metals markets, the London Metal Exchange (LME), in doing so. 'Our goal is to control and stabilise the market price,' explained Megain Widjaja, the young director of ICDX, who felt that the tin price was regularly manipulated. As such, all tin exported by Indonesia would first have to go through the Jakarta stock exchange.

The impact of this policy is still being debated. According to Widjaja, tin prices fluctuated by only 8 per cent, compared to 20 per cent and 30 per cent previously. But according to a London analyst, the prices set by the LME still serve as a benchmark, and he didn't believe that it would change anytime soon.[44] He did concede, however, that 'Indonesia applied a very original idea'. Indeed, to support its industrial policy, the nation needed to build road networks, electricity grids, ports, train stations, and airports. It would take strong and stable mineral prices to pay off and maintain these investments.

So Indonesia warded off the invisible hand of the market by using the controls of the stock market. This inspired other Asian markets. In 2015, the Shanghai Futures Exchange included tin among the metals for trade on its futures market,[45] and announced in 2018 that it would soon include copper.[46] Malaysia followed suit in 2016.[47] Other trading floors also introduced trading platforms for copper, nickel, and zinc.[48]

But Indonesia's nationalisation of its mining resources did not meet with the same success as its Chinese counterpart, for it failed to put the necessary means behind its policy. The colossal investments needed to build the downstream industry were delayed, and the country's trade balance began to teeter as budget shortfalls accumulated. In 2017, the country had no choice but to loosen its strategy by allowing several minerals to be exported again.[49] The drop in commodity prices was largely responsible for this setback, as this nationalistic approach was for the most part implemented during a global 'commodities super cycle' — fifteen years of plenty that began in 2000, during which market prices skyrocketed. It was a boon for buyer countries, but also put seller countries in a position of force that aroused their nationalist instincts.

But the super cycle ended in 2014. The consumer-producer

trade power returned to an even keel, and producer countries began to think twice before investing higher up the value chain. Some say that the old world is resilient, and that it will have the last word. That is, unless emerging countries begin to consume as conspicuously as OECD countries do, giving rise to a new world that will eclipse three centuries of Western order.[50] Yet how do we make ourselves heard by billions of individuals who dream of eating meat at every meal, of drinking champagne, and of travelling to Europe and the US with their families to take a selfie in front of the Eiffel Tower or the Empire State Building?

For emerging countries, rare metals are more than ever a means to a better way of life. It is an unstoppable reality that is only growing: in 1998, after years of struggle, the Kanak separatists of New Caledonia obtained the majority holding in a refinery in the Koniambo massif — the world's biggest nickel reserve. Local populations therefore benefit from local processing of the mineral — synonymous with added value. A similar trend is emerging in Cambodia, Laos, and the Philippines. According to Mostafa Terrab, the CEO of the OCP Group, Africa is also joining the trend: 'There is a growing phosphate-processing industry in Africa to manufacture fertiliser for Africans. It is likely to spread to other sectors. Africa has no choice but to industrialise.'[51]

This intuition is consistent with the 2050 African Mining Vision advocated at the twelfth summit of the African Union in 2009.[52] The aim is to make mines a driver of inclusive growth by capturing a bigger share of their added value. To date, only 15 per cent of African mining production is believed to remain on the continent, so we are far from an optimal result. But given Africa's growing contribution to global GDP, change is inevitable. More importantly, the stakes are no longer purely industrial or political. The fair distribution of minerals — a shared global asset — has become a *moral* imperative, and a cause in which international institutions are now united.[53]

CHAPTER SIX

The day China overtook the West

THESE CONTENTIOUS METALS HAVE LAID BARE A BATTLE OF INVENTION. Today, nations are competing for the brightest minds and most innovative start-ups, and want to claim ownership of the most outstanding patents to epitomise their culture and genius. New technologies promote a certain economic and societal model or lens through which to view the world. China knows this. Through its industrial rare metals strategy, it is betting heavily on scientific development, encouraging the creativity of its people, and ultimately cultivating an alternative civilisation model to the tenets dictated by the West.

Chinese recipes for state-directed high-tech

The theoretical foundations of China's intellectual emulation were laid in 1976, when Deng Xiaoping broke ranks with Mao's agricultural ambitions to proclaim that, from then on, 'production power lay in the sciences'.[1] Subsequent leaders have perpetuated and deepened this conviction: in 2006, President Hu Jintao stated that 'science and technology' form 'the central thread in the development strategy of China';[2] in 2010, the twelfth Five-Year Plan — a road map tracing the key economic strategies for the 2011–2015 period — identified as priorities seven advanced industries and just as many new-technology horizons;[3] five years later, innovation and technological progress formed the centrepiece of the thirteenth Five-Year Plan (2016–2020).[4] Thus, concepts that were never central to China's history had become mantras of the state.[5]

To consolidate this vision, Beijing drew on the terrific competitive advantages of its economy: cheap labour from the inner provinces; cheap capital, attributable to China's policy of devaluing the yuan; and the sheer size of the Chinese domestic market that allows significant economies of scale.[6] Beijing also accelerated competing companies' relocation of their production facilities, using partnership as its weapon. Under so-called joint ventures, foreign businesses gained access to the abovementioned competitive advantages in exchange for their technological know-how and, therefore, their patents. Beijing called this absorption and internalising of foreign technologies 'indigenous innovation'.[7]

The strategy's foundations were set out in an industrial policy document published by the Chinese government in 2006.[8] An

American consultant based in Beijing ironically described the document as 'full of … good intentions and gilded rhetoric about international cooperation and friendship'.[9] The reality is that China's definition of indigenous innovation is reworking and adjusting imported technologies to develop its own technologies. 'The plan is considered by many international technology companies as a blueprint for technology theft on a scale the world has never seen,' a US report published in 2010 asserted. It continued: 'With these indigenous innovation industrial policies, it is very clear that China has switched from defence to offense.'[10]

The Chinese applied this very tactic to rare-earth magnets: it enticed — or forced — foreign businesses onto its territory under the guise of joint ventures, and then launched a process of 'co-innovation' or 're-innovation'. This is how China purloined the technologies of Japanese and US super-magnet manufacturers.

Having reaped the benefits of the invention of others, Beijing built an ecosystem of endogenous creation to 'move from factory to laboratory', starting with a variety of research programs that began in the early 1980s.[11] One of the most emblematic of these, the '863 Program',[12] was launched to make China a leader in seven leading industries, many of which are considered 'green'.[13] The more recent 'Made in China 2025' plan saw the creation of some forty industrial innovation hubs across the country. In 2016 alone, the Chinese state spent around US$520 million on research — less than the US, but more than Europe.[14]

But China has many weaknesses: relative to its population size, it has far fewer researchers than France or the UK; there remain colossal challenges to education; while rural China — a massive part of the country — is sidelined from this momentum. And does the state that has so far managed to combine state interventionism and entrepreneurial freedom run the risk of impeding innovation? More than anywhere else, the success of the innovation ecosystem hinges on administrative efficiency. But achieving this will require the government to commit to painful structural reforms with unclear outcomes. Moreover, the inertia of state-owned enterprises — a formidable part of the energy, telecommunications, and finance sectors — is no longer viable … yet often their leaders are top government officials. How will the government reform these conglomerates without creating tensions and deadlocks within the party?[15]

Lastly, some of China's characteristics do little to aid its cause. While an interventionist regime may have allowed a strategic state to flourish, it leaves no room for any deviation. How can an administration that employs two million government agents to restrict online freedom of expression encourage creativity?[16] A government that stymies the freedom to criticise — and therefore to think differently — nurtures a potent culture of copying, and turns the lack of inventiveness into a building block.[17] 'The Chinese have the technology, but they are stuck in an organisational and intellectual logic that dates back to 1929,' concluded a former Western diplomat posted to Beijing.[18]

Be that as it may, the Chinese authorities are doomed to succeed. Furthermore, innovation, together with the logic of moving upmarket, presents the communist regime with two real, shorter-term challenges:

- The first stems from its strong desire for technological independence, fuelled by a litany of past humiliations. The first dates back to the fallout from the Sino-Soviet split at the end of the 1950s: on the back of diplomatic tensions, in the summer of 1960 the USSR withdrew its technical assistance, on which many heavy Chinese industries depended for their long-term survival.[19] Then, in 1989, in retaliation for the repression of the student movement on Tiananmen Square, the US imposed an embargo on the sale of weapons to China. Beijing learned the painful lesson that it could not rely on anyone's strengths but its own. Since then, an obsession with self-sufficiency has pervaded the Chinese psyche — so much so that China aimed to have decreased its reliance on foreign technologies to 30 per cent by 2020, from 60 per cent in 2006.[20]
- The second challenge is the survival of the Communist Party, which has concluded a tacit contract with one-fifth of humanity: the acceptance of an authoritarian regime in exchange for rising standards of living — a pact that a severe economic slowdown would render null and void. To hold up its end of the bargain, the state has to integrate up to 15 million workers who enter the urban job market every year — an unrealistic feat if China focuses all its efforts on mining alone — indeed, there are more jobs and even greater growth potential downstream of the value chain. Rare metals therefore constitute one of the keys to the resilience of

an authoritarian regime that must constantly innovate to avoid going the way of many imperial dynasties before it.

Astounding technological progress

These factors combined help to explain why the Communist Party has so far been so incredibly successful. 'I was in China ten years ago. Back then, we spoke about textiles, toys, assemblies for electronic products. Quite frankly, no one could have imagined what happened next,' admits a European journalist based in Beijing.[21] China's astounding progress in the electronics, aerospace, transport, biology, machine tools, and information technology sectors caught everyone off guard — including the upper realms of the Communist Party.[22] In aerospace, China has already put a robot on the moon, and it plans to send an astronaut as well by 2036. In 2018 alone it launched some thirty-seven space missions, dethroning Russia as the US's main competitor in the new space race.[23] Beijing wants to move beyond the demand side of new technologies by trading its status of being a skills consumer for that of a skills supplier.[24] In 2018, China filed a staggering 1.4 million patents — more than any other country in the world.[25]

And while we cry over spilt milk, the Middle Kingdom is picking up the pace: it wants to explore the still-unknown properties of rare earths to develop the applications of the future. Some of its university research programs are advanced enough to both astonish and alarm a researcher at the US Department of Defense: 'Losing our supply chain was tragic enough. But now China is busy getting a ten-year head start on us. We could easily find ourselves without the intellectual property rights of the applications of the future that matter the most.' China is openly pursuing this move: 'We want to use these metals to become the world leader in technology,' Vivian Wu said.

On 29 September 2010, Kathleen Dahlkemper, a member of the US Congress, said as much in the House of Representatives during China's embargo of rare earths: 'The Chinese have taken control of the rare-earths market, and we, the United States, are being overtaken.' Her words were crucial: Dahlkemper did not say that the US's technological advance could be reduced or diminished. She did not say that China was catching up with the world's superpower or that the Chinese would put more and more obstacles in its way. She said that we, the West, were about to be overtaken, as is already the case in a growing number of industrial segments. In fact, Beijing has already designed a stealth fighter jet more advanced than that of its Japanese rivals.[26] From 2013 to 2018, the most powerful super computer on the planet came from China.[27] This earned China the title of 'the leading IT power globally'.[28] It has also put into orbit the first quantum communications satellite with reputedly impregnable encryption technology.

Above all, China has taken the lead in an impressive range of green tech. It has put its image as a polluting and polluted country behind it to stand out as the world leader in green-energy generation, the number-one manufacturer of photovoltaic systems, the global hydropower giant, the biggest investor in wind power, and the world's largest market for clean-energy vehicles. China has also started work on a massive network of eco-responsible 'forest cities', including Tianjin, Dongtan Caofeidian, Wuhan, Changxindian,

Taichung, and Hsinchu. Hundreds of eco-suburbs and cities are springing up. In 2018, over US$300 million[29] — one-third of global funding pledged — was invested in these new industries.[30]

This is despite US president Barack Obama's warning while announcing the US's WTO complaint against China's rare earths policy: 'Being able to manufacture advanced batteries and hybrid cars in America is too important for us to stand by and do nothing. We've got to take control of our energy future and we cannot let that energy industry take root in some other country …'[31] The Obama administration looks to have failed in this endeavour: by 2020, China was expected to be producing 80 to 90 per cent of electric vehicle batteries.[32] With a monopoly in the production of rare metals and in the green technologies that depend on them, China intends to become the biggest green-tech-producing country. It wants to siphon green jobs away from Europe, Japan, and the US.

Economically speaking, it wants to be the outright winner of the energy and digital transition.

This ambitious ecological shift will also soothe growing tensions in China over environmental matters that have become unacceptable for the population.[33] Indeed, anti-pollution protests now number between 30,000 and 50,000 every year. Whether protesting against a petrochemical complex project, such as in the city of Kunming, Yunnan, or opposing the construction of a waste incinerator, such as in Hangzhou, Zhejiang, a middle-class movement aligned with the global NIMBY (Not In My Back Yard) trend is openly rejecting the Chinese growth model. There are close to 8,000 environmental non-profits working to harness, coordinate and unify this groundswell.

'The old development model cannot last,' the environmentalist

Ma Jun said to me. 'We cannot continue consuming the way we do. Change is needed.' This green movement is working to modernise the country's growth drivers by giving preference to 'lighter' services and technology, such as digital technologies that have a smaller environmental impact. This will give China a greener image on the global stage, and even place Beijing as the diplomatic leader of the energy transition, thus filling the void left by the withdrawal of the United States from the Paris Accords.

The diminished West

The Chinese strategy of moving up the rare metals value chain was executed at the expense of the industrial vitality of Europe and the US. It implicitly revealed the vulnerability of the Western economic model, despite its establishment as a benchmark at the end of the Second World War.

A German academic put numbers on this reality: looking only at rare earths since 1965, the capture of the oxide market (the powders refined by Rhône-Poulenc) has led to the transfer of US$4 billion of wealth to China.[34] Further up the value chain in the magnet and battery markets, this figure increases tenfold to over US$40 billion. It makes sense: there is greater value for China in capturing the downstream chain.

An Australian researcher applied this reasoning to even more advanced manufacturing sectors, such as the components industry — the assembled parts integrated into a consumer good (such as printed circuit boards, sensors, amplifiers, diodes, LEDs, thermostats, and switches).[35] Following the same logic of the higher capital gain inherent in downstream industries, the transfer of wealth from the rest of the world to China is estimated to have increased tenfold to US$400 billion. The researcher then considered equipment manufacturers — the industries that produce parts that are even more sophisticated than components (in the car sector, this includes dashboards and built-in cameras; in IT, computer hard drives; in aerospace, engines and software for commercial aircraft). The amount of transferred wealth is again multiplied by ten to reach US$4,000 billion — twice France's GDP, or the equivalent of Germany's GDP.

The annual rare-earths market is worth a derisory US$6.5 billion — 276 times smaller than the oil market.[36] But because these small metals are found in just about everything we use, this microscopic industry takes on gigantic proportions. We should also consider that neither of the two studies accounts for the loss of mining industries and end-product manufacturers (such as wind turbines, electric vehicles, and solar panels); the loss of tax revenue for governments; the impact on trade balances; or similar consequences of China moving up the value chain of other rare metals.

And what of the millions of jobs created on one side at the expense of millions lost on the other? In the US, one need only drive an hour out of Chicago along Lake Michigan to the neighbouring state of Indiana to see the disastrous effects of the Chinese battering ram on the US metallurgy industry. In the city of Gary, I met Jim Robinson, a United Steelworkers trade unionist, in his office. His cheerless face told a story of the demise of what was a thriving steel industry until the 1980s. 'The region was so industrialised that it was called the American Ruhr!' he recalled, in reference to Europe's first industrial basin in western Germany. 'It was in its heyday. No one could have imagined what we were in for.' For its residents, Gary is 'a wreck': entire neighbourhoods have been abandoned; houses — their doors hanging off their hinges — are going for US$50. Every day, men and women flee this industrial ghost town ravaged by unemployment, despair, and insecurity.

Three-quarters of magnet manufacturers in the US are gone. The sector that once numbered 6,000 professionals in its ranks in the US now employs no more than 500.[37] Further downstream, Japanese car manufacturer Toyota and its German competitor BMW have also moved some of their operations to China.[38] Japanese conglomerate Sumitomo, the German chemical company BASF, and its companion in misfortune, Grace in the US, have taken the same route. 'Naturally, they were drawn in by China's cheap labour costs. But the biggest push factor for moving their factories was access to rare earths. Millions of jobs were siphoned,' Australian academic Dudley Kingsnorth explained.[39] By betting on renewables, Beijing precipitated the downfall of a fossil-fuel-based industrial order in which the West excelled in favour of a new energy system in which the West is already lagging behind. In a way, one can understand Donald Trump's refusal to commit the US to the energy transition: he would rather preserve an oil-energy model that allowed the US to dominate the twentieth century[40] than commit to going fully electric — a move he knows could be detrimental to American industry.[41]

France was not spared either. Arnaud Montebourg, the then minister of industry, reported as much in his op-ed piece published in a major French newspaper: 'In 2001, a small village in Haute-Garonne, Marignac [in south-western France], painfully witnessed the loss of France's only magnesium factory to Chinese competition. Many years later, with its monopoly on the magnesium market, China increased its production prices to a level that would have returned the French facility to profitability. In the interim, France had lost hundreds of jobs … in car and aircraft lightweighting.'[42]

A lack of interest in the metal-refinement industry also led, in 2013, to the liquidation of the French operations of Comptoir Lyon-Alemand, Louyot et Cie — a company specialised in processing precious metals. 'It was the only French rare metals company,' a former employee said. 'Nearly 4,000 jobs were cut in 2002' in France and globally.[43] The jobs at stake in all the high-tech sectors were highly specialised, requiring advanced expertise. 'They were the "green jobs" that would create a modern and developed economy,' an American trade unionist added.[44] With them went know-how — some of it centuries old — with ramifications for such sectors as defence, electronics, automotive, and, of course, the energies of the future. These human and social dramas add to an already bleak picture: 900,000 industrial jobs lost in France over the last fifteen years, or a 25 per cent drop in the workforce. Over the same period, the secondary sector's contribution to France's GDP slid four points.[45] The figures are not much better in Europe and the US, where the industry made up 22 per cent and 18 per cent of GDP respectively in 2018, down 11 per cent and 12 per cent since the turn of the century.[46]

De-industrialisation in the United States and Europe hammered the post-war social contract, stirring up deep social turmoil and creating a hotbed for a franchise of populist parties. Donald Trump succeeded in reaching the White House because he could count on the voters in the de-industrialised states of the Rust Belt. In these swing states, where votes can tip the result of a national election, the Republican candidate vigorously denounced the anti-competitive practices of the Chinese and offshoring, and emphasised the need to protect the US from the industrial war spearheaded by Beijing. The strategy paid off, and Trump won the popular and Electoral College vote in virtually all these states, wiping out the comfortable popular-vote majority enjoyed by Hillary Clinton at the national level.

The West would have felt the pernicious effects of de-industrialisation with or without rare metals. But this economic, social, and political crisis was amplified by China's monopoly of the resources destined to replace fossil fuels, and by its formidable strategy of absorbing the green industries reliant on rare metals. Moreover, the European model has proven 'powerless to implement a policy that preserves its economic, technological, and social assets', writes a Western expert, adding: 'The survival … of European democracy … may be the last hurdle in the way of the scarcely begun emergence of Chinese industry.'[47]

When two world views collide

Meanwhile, China's success has allowed it to foster a model of government that values a long game, unlike the short-sighted decisions of the West that have wrecked most of its industrial policies. This 'authoritarian capitalism … provides encouragement to other autocratic states', suggests an Indian academic.[48] It has proved it can produce solid growth while guaranteeing political stability, thereby giving substance to the 'Beijing consensus' — the idea that the Chinese development model can serve as a benchmark for other emerging countries.[49]

This consensus opposes another, that of Washington, which has prevailed since the end of the Cold War, and by virtue of which economic growth and democratic progress are necessarily correlated. The war for rare metals — and green jobs — has therefore lifted the lid on a new ideological conflict that pits China's principles of political organisation against those of the West.

'A clash of civilisations is a very Western way of seeing things!' says Zhao Tingyang. Zhao is a philosopher who became famous in China by popularising *Tianxia*, a concept inspired by the teachings of Confucius, which promotes the search for harmony in international relations.[50] He had agreed to talk to me about the future of China's relationship with the West. He said that the centrifugal force of globalisation 'pushes us to become more and more interdependent, making conflict — military or economic — an undesirable option for everyone'. In this world at peace, the philosopher predicted, *Tianxia* will reign; it will be a new era of power led by the globalisation of communication and transport, which will unite a cosmopolitan ruling elite and share Western and Chinese values. 'I believe this system will create peaceful interactions and better common understanding,' he said.

At a large table laden with fine food, I met with a Chinese rare metals heavyweight who concurred. 'The world to come is more open and cooperative,' he said in between chopstick manoeuvres.[51] A corollary to this feel-good statement is the serenity of China's 'panda diplomacy', which involves gifting specimens of these giant folivores to countries with which it wishes to forge diplomatic ties. You would have to be heartless not to fawn over online videos of China's favourite furry mammal rolling about and eating bamboo. With the panda, China seeks to convey the idea of a peaceful emergence, at all costs.

It's tempting to believe in this, as it is to dream of an ethnically diverse world at peace. Except when the game is given away by one detail: by acquiring a strategic rare-earths company in Indiana from under America's nose, China has whipped the covers off its impressive military program.

CHAPTER SEVEN

The race for precision-guided missiles

HOLLYWOOD HAS TAKEN A FERVENT INTEREST IN RARE EARTHS. AND rightly so: with plots featuring rare resources as indispensable as oil, the threat from China, and the survival of high-tech industries, it's blockbuster material. This certainly wasn't lost on the screenwriters of *House of Cards*, the hit television series about Frank Underwood (played by Kevin Spacey), an American politician who ruthlessly climbs his way to power in the White House. Rare metals are woven into the storyline in the second season, which bears a few similarities with reality.

In one episode, China uses its 95 per cent monopoly of the global production of samarium-149 — portrayed as a very rare metal that is indispensable to the operation of American nuclear reactors — to make the US pay a hefty price for the resource.[1] Electricity prices become prohibitive for consumers, and Washington is plunged into political disarray. On Frank Underwood's suggestion, the US decides to bypass the Chinese monopoly by buying samarium through a third party. This forces the Chinese, says the protagonist, 'to lower [their prices] to keep a direct flow with us. We stockpile samarium for defence purposes … and quietly sell off what we don't need to our friends …'

For the Pentagon, there is nothing new about stockpiling rare metals needed for the country's defence as vital components of its war arsenal: tanks, destroyers, radars, smart bombs, antipersonnel mines, night-vision equipment, sonars, or the new laser weapons already tested in the Persian Gulf.[2]

More pressing still is the growing strategic importance of these resources as the threat of cyber warfare escalates. Indeed, twenty-first-century warfare is taking place on more and more fronts: belligerents strike not only on land, but in the air, in space, in cyberspace, and through media channels by seeking to wipe out the enemy's communication channels, control images, rewrite history, and manipulate opinions. Warfare has moved from the ground to the stratospheres of electronic, media, and virtual wars, with rare metal–hungry servers, drones, radar aircraft, constellations of satellites, and space launchers as arsenal.[3] The further away we move from the battlefield, the deeper we need to dig below it.

In terms of physical volumes, armies do not need a great deal of rare metals. According to one expert, the US defence industry imports a total of 200 tonnes of magnets every year — just 0.25 per cent of global production.[4] A London analyst reckoned that the rare earths required to meet US defence needs for three years could fit into a single backpack.[5] Yet the world's powerhouse, whose military under Donald Trump was expected to monopolise US$738 billion of the US budget in 2020, would be cut down to size should these few packages of rare metals not make it to the arms factories.[6]

Pooch parlours and precision-guided missiles

For decades, the Pentagon sourced its magnets from manufacturers in the US. One of the most strategic of those was a company called Magnequench. By specialist accounts, it was the best rare-earth magnets producer in the world: its factories were the pinnacle of the production line for Abrams battle tanks and for Boeing's JDAM smart bombs, which were used in the Afghanistan and Iraq wars.

The highly prized supplier of the US Army ran its successful outfit out of Valparaiso — a charmless town of 32,000 people in the state of Indiana, a two-hour drive from Chicago. Former Magnequench employee Terry Luna agreed to escort me to what was once the manufacturing area. Although somewhat worse for the wear, the buildings, tall and rigid in the humid summer heat, were unchanged. But their occupants were different. 'I started at Magnequench at the front desk,' explained the full-figured woman in a nasal voice. 'But if you look at the sign, it says, "Coco's Canine Cabana": a doggy day-care centre.'

We went in and were given a warm welcome by the centre's manager, a young blonde woman wearing a T-shirt and ripped jeans. We walked across a hall, since repurposed for holding birthday parties for pooches, and through a grooming parlour complete with doggy shampoo and clippers — 'made in China' (of course). We then reached the central storage area. 'Here they kept magnets for bombs, remote-controlled missiles, and a whole lot of war weapons,' explained Terry, barely holding back her tears. 'And they sold the lot! It all went to the dogs!' In 2006, Magnequench closed its strategic plant in Valparaiso to open a new one — in Tianjin, 130 kilometres south-east of Beijing.[7] 'The Chinese learned our industrial secrets,' added Terry. Two hundred and twenty-five employees were laid off. It's a tragic snapshot of an America whose most qualified workers abandoned their manufacturing tools to make way for pets.

Magnequench's fate paints a picture of Beijing's new military ambitions while it imposes its status as an industrial power. China's defence budget is now second to that of the US, spending US$177 billion in 2019 compared to US$123 billion in 2010.[8] China's goal? To go as far as dethroning the US by 2049 — the year that the People's Republic of China celebrates the centenary of its founding.

Since the 1980s, the Chinese army has developed in three ways. First, it changed its doctrine when Deng Xiaoping abandoned the dogma of 'the people's war', according to which battles took place in the heart of the Chinese territory. This was replaced with 'the people's war under modern conditions', whereby armed forces fight at the country's borders and beyond. The next change was organisational, as the army moved from the idea of 'the power of many' to a smaller, specialist army. The last change was technological, after the Gulf War left China facing the reality that it needed to catch up with the US.[9] To achieve this third transformation, Beijing undoubtedly gave much reflection to history's golden rule about natural resources: peace and metals rarely make good bedfellows.

Six thousand years ago, when humans began to melt copper, they were able to put aside their stone tools for sharper and more solid implements. At best, humans could perfect their hunting techniques with this breakthrough until 2,000 years later, when the Sumerians discovered bronze — a robust alloy of copper and tin. This time, empires and civilisations were emboldened to turn out swords, knives, and axes, raise armies, and enter what would be the very first arms race in history.[10]

Around the twelfth century BC, in the south of modern-day Turkey, the Hittites melted an even lighter and more widely available metal — iron — to forge weapons that were more powerful and easier to wield. This, say some historians, led ultimately to the European conquest of the Americas.[11]

Then came steel, which in 1914 tipped Europe into an industrial war. The iron and carbon alloy was used to make shell casings, the first modern fragmentation grenades, hardier helmets for soldiers, and armoured tanks — all of which contributed to the bloodbath that was the First World War.

Every time a people, civilisation, or state masters a new metal, it leads to exponential technical and military progress — and deadlier conflicts. Now it is rare metals, and in particular rare earths, that are changing the face of modern warfare. China knows that the master of their production and application will undeniably have the upper hand strategically and militarily. Targeting Magnequench and acquiring its patents and, therefore, its secrets was a perfectly logical move.

But the Magnequench affair gave rise to grave national-security concerns, as exposed to the public in 2015 by the CBS television program *60 Minutes*. All it took was the factory's removal from US soil to put the world's military powerhouse at the mercy of Beijing for the supply of some of the most strategic components of its military technology. How could America have put itself in such a precarious position?

Magnequench in the radar of the 'Red Princelings'

'I spent twenty-one years in the office of the Secretary of Defense, working on tech transfer … and in those twenty-one years, the Magnequench case ranks among the top five of the thousands of cases that I've seen.' Peter Leitner was a senior official at the US Department of Defense at the time of the Magnequench affair. His job was to review all technology exports that might have a negative impact on America's military sovereignty, and to stop the export if required. The buyout of Magnequench was in fact already underway in the 1990s, during the presidency of Bill Clinton — years before the factory was moved. Magnequench's parent company, General Motors, agreed to sell the magnet manufacturer to the Chinese in exchange for approval to build a vehicle plant in Shanghai.[12]

During that time, Leitner and his colleagues were caught up in another matter: the sale by Kentucky-based US company Crucible Materials of its rare-earth magnet-production operations to YBM Magnex International, a company listed on the Alberta stock exchange. On paper, YBM seemed like a perfectly above-board business, with a head office and warehouse in Philadelphia. Upon closer inspection, however, it turned out to be a shell company owned by the 'Red Mafia' — a Russian criminal network thriving in the void created by the fall of the Soviet Union.

YBM was, in fact, hand in glove with a mob of shady businessmen, including a Ukrainian national by the name of

Semion Mogilevich, whom the FBI would later describe as 'a global con artist and ruthless criminal … involved in weapons trafficking, contract murders, extortion, drug trafficking, and prostitution on an international scale'.[13] Under the guise of selling magnets, the company busied itself with laundering the proceeds of criminal activities perpetrated in Russia and in numerous other states of the former Eastern bloc.

Crucible Materials nevertheless sold its magnet operations to YBM on 22 August 1997, from which time a part of America's production of the magnets needed for its defence was at the mercy of organised crime. By putting itself in the running for an industry as strategic as this one, did YBM make use of a geopolitical agenda dictated at a higher level to serve its own interests? Russia's state of decline at that time hardly put it in a position to oversee such an operation, and the motives of the key players in the affair are vague to this day.

The stakeholders of the Magnequench buyout proved to be just as dubious. Meet Archibald Cox Jr, the man who bought Magnequench. The son of a special prosecutor during the Watergate scandal and the chairman of the private equity firm Sextant Group, he was also a seasoned trader who was drawn in by the alluring operation.[14] 'He was a very slick figure,' recalled Leitner. 'He wasn't shy about charming those in charge at the Department of Defense who had concerns about the deal. He'd say, "Don't call me Cox, call me Archie!" They were seduced by his charm, and were at that point predisposed to approve anything he wanted. It was just incredible.'

But Peter Leitner and his colleagues were fast off the mark; they knew Archibald Cox and the Sextant Group were the screen between Washington and Beijing. Through a subtle misdirection of funds into tax havens in the Caribbean, dubious Chinese businessmen had been brought into the deal. 'One of them was the "Red Princeling" [the heir to one of the top officials of the Chinese Communist Party]. It was none other than Deng Xiaoping's son-in-law.'

It is public knowledge that Deng Xiaoping handed the reins of the China Nonferrous Metal Mining Group Corporation (CNMC) — a formidable state-owned mining company — to his son-in-law Wu Jianchang to ensure the longevity of the strategic nonferrous metal sector.[15] It was the very same Wu Jianchang who, through CNMC's New York branch, was directly involved in the sale of Magnequench.

Peter Leiter later discovered that not one, but *two* of Deng Xiaoping's sons-in-law were involved. Zhang Hong, married to Xiaoping's daughter Deng Nan, was chairman of Beijing San Huan New Materials High Tech Inc. — Magnequench's final buyer. This wasn't just any buyout of an American company. 'All these things were riddled with ambiguity,' said the former civil servant, who reported the problem to his superiors. 'We pointed this out repeatedly. We went to our superiors within the defense department, we argued ad infinitum, and eventually we were ignored.'

The apathy of the Democratic administration is even more surprising given that China was in the process of executing its 'Sixteen-Character Policy', which it had outlined as early as 1978.[16] Enacted by Deng Xiaoping, the strategy included a program to acquire 'dual-use' technologies to ultimately strengthen China's army. Magnequench was an ideal choice, as its magnets were used for the vehicles of General Motors and for the US Army.

The premise of the Sixteen-Character Policy was pragmatic: given the difficulty in procuring war technologies due to the US arms embargo, China would buy foreign companies whose know-how in civil applications could be repurposed for more hostile ends. In the years that followed, this strategy would lead to an extraordinary proliferation of Chinese espionage against the US. According to a former US counterintelligence agent, 'China's intelligence services are among the most aggressive [in the world] at spying on the US.'[17] A European researcher explained that Beijing's interest was in two technologies in particular: those used in network-centric warfare, allowing armies to use information systems to their advantage; and smart bombs, containing the very magnets produced by Magnequench.[18]

China's attempt to influence US ballot boxes

Any US politician who was aware of Beijing's intentions and actions knew full well that a company such as Magnequench would be a choice target for the Chinese military. One such politician was George W. Bush, who, as president of the United States, could have stopped the Valparaiso plant from going to China in 2006. But the US was busy leading a global war against terrorism; any threat not linked to Islamic fundamentalism stood little chance of capturing the attention of the White House. During the Democratic Party's Indiana presidential primary in 2008, in which Barack Obama was narrowly defeated, Hillary Clinton stated that Bush 'did nothing' to stop the demise of Magnequench. Rare metals experts quickly pointed out the hypocrisy of the claim — wasn't it the party's candidate's husband who, only a few years earlier, had allowed the negotiations that led to the buyout, despite warnings from numerous top officials at the Department of Defense?

Why was there such apathy on the part of the Democratic administration? This is where the case of Magnequench takes a troubling turn. At the time of these events, a series of technology transfers had already started to raise the hackles of the US military industry and the small circle of magnet manufacturers. One of these was Steve Constantinides. He later described the stupefaction of one of his peers upon learning that the White House had, over a period of three or four years, been divulging confidential industrial intelligence on American ballistic technologies to Beijing: 'The United States shared its industrial secrets on missile technologies with China. And Bill Clinton forced his government to do so.'[19]

What was happening behind the scenes? 'Everyone had their reasons,' replied Constantinides evasively, not wanting to say any more on the matter.[20] Leitner was more forthcoming: 'Some people argued it was because of the kickbacks and money that was given to the Clintons, the Republicans, and the Democrats … typically from the People's Liberation Army — money from the Chinese that went directly into the White House.' But without any real evidence, such claims remain nothing more than rumours.

Subsequent documentaries co-directed by Leitner investigated the clandestine relationship between the Democratic administration and China in the 1990s. It is widely known that Beijing sought to fund the Democratic Party during the 1996 presidential elections in order to back Bill Clinton and his vice-presidential candidate Al

Gore. Despite a US electoral law prohibiting non-American citizens from financially participating in the election process, numerous intermediaries contributed on behalf of China.

Enter the enigmatic Johnny Chung. The Sino-American citizen was so close to the Clinton couple, and his dealings so dubious, that he became known as the Democrats' key channel of funds from China. It has since been revealed that Chung gave the White House funds that he received indirectly by top officials in the Chinese army.

These facts were reported by the US press and in a documentary in which Peter Leitner was interviewed.[21] James Woolsey, the head of the CIA at the time, also agreed to be interviewed by the director, who asked him why he'd only met with Bill Clinton twice over two years, while Johnny Chung went to the White House fifty-eight times over the same period. 'Mr Woolsey could not explain it,' recalled Leitner. 'He just said that the president's schedule speaks for itself.' Chung allegedly went as far as arranging a face-to-face meeting for a high-ranking official of the People's Liberation Army — the source of the financing — with the president during a fundraising evening in Los Angeles.[22] However, the justice system quickly caught wind of the affair.[23] *The Washington Post* leaked a federal enquiry into the Chinese embassy in Washington's heavy-handed coordination of China's interference efforts. The Democratic Party was eventually ordered to return millions of dollars in donations to their many benefactors.[24]

Dozens of people, including the bagman Johnny Chung, were convicted in the scandal that became known as 'China Gate'. Yet the senior level of the Democratic Party was never targeted, and the attorney-general, Janet Reno, refused to appoint an independent counsel to take the investigation further. Interestingly, when President Donald Trump was faced with an in-depth enquiry into his alleged ties with Russia during the 2016 campaign, no one seemed to remember the other proven scandal involving a country infinitely more harmful to Washington's interests.

Is it possible that the clandestine funding received by the Democratic camp pushed President Clinton to divulge US technology intelligence in return? It's a serious allegation. Did the White House think that the technology in question was not as strategic as experts claimed it to be, or that, with or without Magnequench's patents, the Chinese would have acquired the technology anyway? There are many grey areas that may never be brought to light. Either way, as stated by Peter Leitner, 'The Clinton administration was heavily predisposed toward approving almost anything the Chinese wanted.' Leitner also left me with an open question — one that would take months of investigation to clear up: were rare-earth magnets the essential thread in this tangle of corruption, cynicism, and a thirst for power?

South China Sea: access denied

Peter Leitner did make one thing clear: 'The Chinese targeted Magnequench in order to advance their ability to produce long-range cruise missiles.' The question is whether the technology they acquired formed part of the two ballistic missiles that were first revealed to the world during a spectacular military parade in Beijing in 2015: the Dongfeng-26 ballistic missile, capable of reaching the

US base on the island of Guam, and the Dongfeng-21D anti-ship ballistic missile.

Nicknamed the 'aircraft carrier killer' and operational since 2010, the DF-21D has been central to Beijing's policy of prohibiting access to the South China Sea these past few years. Having control over this strip of ocean running from its coasts to the south of Vietnam would increase China's strategic leverage, and give it access to prodigious quantities of offshore hydrocarbon resources, as well as an eye on the comings and goings of half the world's oil.

This scenario is unacceptable to Japan, South Korea, Vietnam, and the Philippines, but especially to the US, which several years ago planned to position 60 per cent of its warships in the Pacific by 2020.[25] Barely a week goes by without a naval incident of some sort, making the territory the powder keg that could ignite a Sino-American conflict.

The West's global military supremacy is indisputable. The US accounts for 36 per cent of the world's defence expenditure, compared to China's 14 per cent.[26] And yet Beijing's capability in advanced ballistic technologies has already shifted the balance of power in the South China Sea. 'In a future conflict situation, our military is going to face an enemy that is a lot more robust, that is a lot better armed, is much more modern and technologically sophisticated, and has a great deal of more precision in their targeting capabilities than they should have,' predicted Peter Leitner. 'It means that we're going to be pushed further and further out into the Pacific.'

China is the world's second-biggest economic power, and its relationship with the US has deteriorated considerably. President

Donald Trump has engaged in a fierce trade conflict with Beijing, and his entourage foresees war with the Middle Kingdom over the South China Sea.[27] Wouldn't the US be vulnerable against an adversary that is also the source of its most critical defence components? And would China not take timely advantage of this dependence, either by playing the rare-earths card during trade negotiations, or by hampering America's military efforts?

During questioning before the Senate Intelligence Committee in 2017, CIA director Mike Pompeo stated that the United States' dependence on rare-earth supplies from China remains 'a very real concern' for the agency.[28] He also made sure to expand on the concrete action the CIA was undertaking to address the issue. In 2014, Anthony Marchese, chairman of the mining company Texas Mineral Resources, posed the question directly to Michael Morell, the CIA's deputy director from 2010 to 2013: 'Morrell said to me: "The good news is that the matter is in my inbox. The CIA is perfectly aware of the problem. The bad news is that it's right at the bottom of my inbox, because the White House has never indicated rare earths as a priority to the agency."'

Yet political interest was revived on 21 July 2017, when President Trump ordered a report to assess and strengthen 'the manufacturing and defense industrial base and supply chain resiliency of the United States'.[29] Delivered in the spring of 2018, the report provides a database of the American industries that guarantee the country's military sovereignty. More importantly, it lists all the 'individual points of failure': the critical companies and factories whose extinction would paralyse the entire US defence industrial base.

Anthony Marchese again decided to dig deeper. During the same period, he set up a meeting with one of the report's authors at a coffee shop outside the White House. He learned that 'the White House doesn't like the idea that the Chinese supplies the US its rare earths. So, the administration is thinking of introducing a "Buy American Clause" in its defence contracts.' The protectionist clause would force US defence groups to source their military components from US suppliers, which would naturally revive the domestic production and transformation of rare earths.[30]

Commerce secretary Wilbur Ross also announced his intention to restrict Chinese aluminium imports — far from being used for soda cans, aluminium is also a component of numerous American weapons: 'At the very same time that our military is needing more and more of the very high-quality aluminum, we're producing less and less of everything, and only have one producer of aerospace-quality aluminum.'[31]

The US Department of the Interior has identified no less than thirty-five minerals considered critical to the country's national security and economy.[32] Lisa Murkowski, a senior Republican senator from the state of Alaska, is fiercely committed to the cause, and moved to legislate the American Mineral Security Act: 'Our reliance on China … for critical minerals costs us jobs, weakens our economic competitiveness, and leaves us at a geopolitical disadvantage.'[33] The reality, however, according to an American expert, is that, '[d]espite Mr Trump's rhetoric, to date very little has changed'.[34]

The fact is that the US administration is divided on the issue. Advocates for the status quo say that the cost of regaining America's sovereignty for its rare-earth supplies is too high for volumes so low, that the Pentagon will always find a way to procure the resources it needs (including from the black market), that there is no threat to the United States' military superiority, and that, in any case, the Chinese wouldn't dare trigger a large-scale crisis by trying to threaten supplies to the US Army.

But if the US does not regain its mineral sovereignty, it will, like an unwitting player in a game of Go, allow Beijing to surround its pieces on the board. 'If we leave the Chinese unchecked,' warned Marchese, 'the United States will lose the military status it currently enjoys.'

Some say that the time has already come — as current events would seem to suggest. In spring 2019, no sooner had Donald Trump banned the Chinese giant Huawei from selling to the American telecoms market than Xi Jinping, together with the vice-premier in charge of economic affairs, Liu He, was photographed on a plant tour of JL Mag RareEarth, a rare-earths magnet producer based in the southern province of Jiangxi.[35] No statements were made, but the message was crystal clear: should the trade conflict between the two world powers continue to escalate, Beijing could retaliate by suspending rare-earth exports to its rival. The message was made plain in black and white by the state press agency *New China News Agency*: 'By waging a trade war against China, the United States risks losing the supply of materials that are vital to sustaining its technological strength.'[36] Tensions with Beijing are so high, and the vulnerability of the US so great, that the situation is no longer politically tenable. The sovereigntist approach of the Trump administration, say some, can do little more than loosen China's stranglehold on rare metals production.

In fact, following the threat of embargos by China in spring 2019, the US government published another report calling for the country's reliance on Chinese supplies to be reduced by reopening rare-earth mines, through recycling, and by finding new substitutes. Supporting this, again, was the secretary of commerce, Wilbur Ross. 'The federal government will take unprecedented action to ensure that the United States will not be cut off from these vital materials,' he said.[37]

This gradual awakening by the US is confirmed by a number of industrialist initiatives. The mining company Molycorp, bought in 2017 by the US-based consortium MP Mine Operations LLC, resumed operations, and in 2018 produced 12,300 tonnes of rare-earth concentrate.[38] But if the US has to send its rare earths to China to turn them into metal, these efforts will be in vain. The key issue is refining. This is why in 2019 the US company MP Materials announced that from the end of 2020 it wanted to do its own refining in the US: over 5,000 tonnes per year of neodymium and praseodymium.[39] Washington may also get the nod from Australian mining house Lynas Corp, the largest non-Chinese producer of rare earths. In May 2019, it announced that, together with Blue Line Corp, it would soon be building a heavy rare-earths separation facility in Texas.[40]

Until these announcements become reality, Washington can expect the same difficulties that have created thorny dilemmas for Democratic and Republican administrations — the same dilemmas, in fact, faced by the manufacturers of the revolutionary Lockheed Martin F-35 Lightning II fighter jet.

Fatal attraction: Chinese magnets and the Pentagon

The case of Lockheed Martin finds its origins in a 1973 US law that banned the procurement from foreign suppliers of specialty metals for use in military technology.[41] Lawmakers recognised that parts containing cobalt, zirconium, and titanium were increasingly important to the US Army's arsenal. America needed a domestic industrial base that could ensure the country's mineral security in times of war.

In the early 1990s, the world's leading army undertook a formidable challenge: to design and put into the sky an aircraft that could rival the French Rafale. Developed by the US defence group Lockheed Martin and co-financed by numerous US allies, the F-35 fifth-generation stealth fighter jet has already cost US$400 billion, making it one of the most expensive programs ever run by the US army. America's hopes in the aircraft are as high as the tax bill for citizens: not only will the F-35 allow the US to dominate the sky, it will stimulate the country's defence industry, restore the trade balance, and create tens of thousands of jobs. Over the next few decades, some 2,500 orders for the F-35 are expected to be delivered to Australia, the UK, the Netherlands, Israel, Italy, Turkey, Japan, and South Korea.

Then, in August 2012, two of Lockheed Martin's biggest suppliers, Northrop Grumman and Honeywell, went to the White House with their concerns about the rare-earth magnets used in some of the radars, landing gears, and computer systems that they supplied for the assembly of the F-35. Northrop Grumman had discovered that its radars, installed on the 115 already-delivered stealth jets, contained magnets that were not made by a US manufacturer, but by a Chinese competitor, ChengDu Magnetic Material Science & Technology. Apparently, an unscrupulous intermediary had circumvented US regulations, making the F-35 program partially unlawful and the continued purchase of these parts out of the question.

The Pentagon was alerted to the matter, which was taken up by Frank Kendall, the undersecretary of defence for acquisition, technology, and logistics. It was a dilemma: waiting for an American magnet manufacturer to provide the precious parts to replace those of the Chinese would potentially lead to delays in the F-35's deployment, and the cost of retrofitting the stealth jets with American parts to replace the offending magnets would be astronomical.

The Pentagon therefore looked into waiving the 1973 law for matters of national security. But some top officials were circumspect: how could they be sure that the rare-earth magnets China supplied to F-35 manufacturers didn't also contain spyware? Could the US$400 billion program be 'contaminated' by a few magnets costing no more than $2 a piece? By giving the Chinese control of the downstream supply of rare minerals, had the US given its competitors the opportunity to steal their military secrets and make up for lost ground?

These questions raised another broader question of national security that the US has asked itself time and again: how does it prevent the infiltration of Trojan horses in the microchips and other semi-finished goods containing rare metals sold by the Chinese around the world, including to Western armies? A 2005 report by the Pentagon even raised the possibility of electronic systems that are used extensively in US weapons being infected by malware that could disrupt combat equipment mid-operation.[42] These fears escalated when the Pentagon discovered that raw materials from China were being used in other vitally important military equipment — namely, Boeing's Rockwell B-1 Lancer long-range bomber, certain Lockheed Martin F-16 fighter jets, and all the new SM-3 Block IIA defence missiles manufactured by Raytheon.

Kendall ordered Lockheed Martin to find a solution. Until then, replacing the magnets one by one was out of the question. The technological superiority of the world's leading army and of numerous Western allies was at stake at a time when China and Russia were developing their own stealth fighter jets. Faced with severe time and budgetary constraints, and ruling out the risk of components containing backdoor technology, Kendall made his decision: the ban imposed by the law of 1973 would not apply to some of the rare-earth magnets produced by the ChengDu Magnetic Material Science & Technology Co., making the Chinese company the official supplier of the F-35.[43]

The United States cannot do without Chinese magnets, and so, to this day, Anthony Marchese explained, the Pentagon continues to grant the waiver. 'The manufacturers of the F-35 still buy rare earths in China. Period.'

CHAPTER EIGHT

Mining goes global

DIGITAL TECHNOLOGIES, THE KNOWLEDGE ECONOMY, GREEN energies, electricity logistics and storage, and the new industries of space and defence are diversifying and expanding our need for rare metals exponentially. Not a day goes by that we don't discover a new miracle property of a rare metal, or unprecedented ways of applying it. Indeed, our technological ambitions and dreams of a greener world are limited only by the bounds of our imagination. This means expanding our mining operations across the entire planet — and the Earth, we tell ourselves, will keep up. Contrary to speculation, there will always be that acre of mountain, that crease in a hill, or that clearing in a valley where we can extract those few aggregates of precious particles — that 20-gram hit of rare earths needed by every human on the planet every year.

After all, there is a precedent: between the end of the First World War and 2007, the annual production of fourteen of the minerals essential to the global economy increased twentyfold.[1] By the end of the Second World War, consumption began to skyrocket, taking with it just about every indicator: life expectancy, consumer habits, the accumulation of wealth, the number of possessions, the quantity of electronic data movements, and global warming.

Where does that leave us for the century ahead? Will this breakneck pace simply gather momentum? If global GDP continues to grow at an annual rate of 3 per cent, as it has for the last twenty years, it will have doubled by 2041. By the time you read these lines, everything we build, consume, barter, and throw away will have doubled in less than a generation. There will be twice as many high-rises, highway interchanges, chain restaurants, industrial livestock-production farms, commercial aircraft, e-waste dumps, and data centres. There will be twice as many cars, connected objects, refrigerators, barbed-wire fences, and lightning conductors.

We are going to need twice as many rare metals.

A metals shortage ahead?

Some have put numbers on our future needs. At a symposium held at Le Bourget in 2015, on the margins of the Paris climate talks, a handful of experts presented their forecasts.[2] They predicted that by 2040 we will need to mine three times more rare earths, five times more tellurium, twelve times more cobalt, and sixteen times more lithium than today. Olivier Vidal, a researcher at the French National Centre for Scientific Research (CNRS), even conducted a study of the metals we need in the medium term to sustain our high-tech lifestyles.[3] His work was published in 2015 and mentioned on the BBC.[4] He has given some thirty lectures in Europe, mostly to students. And it stops there.

Yet Vidal's study should be on the bedside stand of every head of state the world over. Basing his research on the most widely accepted growth outlooks, he highlights the massive quantities of base metals we will need to extract from the subsoil to continue to fight against global warming. Take the case of wind turbines: by 2050, keeping up with market growth will take '3,200 million tonnes of steel, 310 million tonnes of aluminium, and 40 million tons of copper'.[5] Indeed, wind turbines guzzle more raw materials than previous technologies: 'For an equivalent installed capacity, solar and wind facilities require up to 15 times more concrete, 90 times more aluminium, and 50 times more iron, copper, and glass than fossil fuels or nuclear energy.'[6] According to the World Bank, which carried out its own study in 2017, the same applies to solar and hydrogen electricity systems, which 'are in fact significantly *more* material intensive in their composition than current traditional fossil-fuel-based energy supply systems'.[7]

The overall conclusion is aberrant. Because global metal consumption is growing at a rate of 3 to 5 per cent per year, '[t]o meet global needs by 2050, we will have to extract more metals from the subsoil than humanity has extracted since its origin'. This bears repeating: over the next generation, we will consume more minerals than in the last 70,000 years, or five hundred generations before us. Our 7.5 billion contemporaries will absorb more mineral resources than the 108 billion humans who have walked the Earth to date.[8]

Vidal admits that the study is incomplete: assessing the actual ecological footprint of the green transition requires a far more holistic approach that includes the life cycle of raw materials. It also requires measuring the staggering volumes of water consumed by the mining industry, the carbon dioxide emissions produced by the transportation, storage, and use of energy, the still little-known impact of recycling green technologies, and all the ways in which these activities pollute ecosystems — not to mention the extent of their impact on biodiversity.

'It's mind-boggling,' admitted the researcher.[9] And yet so few political leaders truly grasp all these aspects. Vidal maintains that in recent years he tried to alert the French minister of research: 'I never made it past the first barriers of the lower administrative hierarchy.' His disappointment was shared by Alain Liger, who held a symposium on rare metals during COP 21. 'I sent a note to Ségolène Royal [the then French environment minister], Emmanuel Macron [the then French finance minister], and Laurent Fabius [the then French foreign affairs minister]. Macron's staff called to congratulate me on the event. But I didn't hear back from Fabius or Royal' — the very ministers leading the climate talks.[10]

Clearly, scarcity is an issue. On the one hand, advocates of the energy transition are adamant that we can draw infinitely on the inexhaustible sources of energy generated by the tides, the wind, and the sun to make our green technologies work. On the other hand, rare metal hunters warn that we could soon run out of several raw materials. Just as we have a list of threatened animal and plant species, we may soon have a red list of metals nearing depletion. At the current rate of production, we run the risk of exhausting the viable reserves of fifteen or so base and rare metals in under fifty years; we can expect the same for five additional metals (including currently abundant iron) before the end of the century.[11] In the short to medium term, we are also looking at potential shortages in vanadium, dysprosium, terbium, europium, and neodymium.[12] Titanium and indium are also stretched, and cobalt is heading in the same direction. 'This will be the next metal shortage,' predicted one expert. 'No one saw this coming, and time is running out.'[13] (See Appendix 14 for the summary table of the viable reserves lifespan for the main metals needed for the energy transition.)

Will we manage to start up enough mines in the next thirty years to satisfy our appetite for metals? What if climate change drastically reduces the water reserves needed to extract and refine minerals? Will we have come up with the technology that will enable us to access poorer, less accessible, and deeper ores when the more abundant mines are depleted? Our era is often referred to as the 'New Renaissance': we are at the dawn of an age of unprecedented technical invention and opportunities for exploration. Yet how can we hope to reach these new frontiers if we run out of the resources we need? What if Christopher Columbus, with no wood at his disposal, hadn't found the *Pinta* and the *Niña* caravels moored at an Andalusian port in 1492?

The energy and digital transition at stake

Diplomatic achievements, ambitious energy-transition laws, and the efforts of the most zealous environmental defenders will amount to nought without sufficient quantities of rare metals. If current data is anything to go by, the green revolution will take much longer than hoped. More importantly, it will be a green revolution led by China — one of the few countries with an adequate supply strategy — and Beijing will not go out of its way to increase its rare metals output to meet the needs of the rest of the world. Not only because its trade policy puts the West at its mercy, but also because China worries about running out of resources too quickly. The black market for rare earths, which caters for one-third of official demand, is using up mines much faster, with some reserves facing depletion from 2027.[14]

Containing this increase in the mining of certain rare metals is vitally important. That is why China is ready to stockpile what it produces — for itself. It already consumes three-quarters of the rare earths it extracts — despite being the sole supplier — and, given its appetite, it may well use up all of its rare earths by 2025 to 2030.[15] The output of any of China's future rare metals mines inside or outside its borders will not go to the highest bidder, but will be taken off the market and channelled to Chinese clients only. And these resources will be snapped up, irrespective of the price tag. 'What will be left over for the rest of the world?' asked an American expert. 'The answer is: nothing. Absolutely nothing.'[16] Beijing will focus on the interests of its green-tech businesses, and support the growth of its energy and digital transition to the detriment of other countries.

This is how China can spurn all stereotypes of being one of the most polluted countries on the planet to instead being seen as spearheading a greener world and the fight against global warming. It would be plausible for three reasons:

- First and foremost, our collective denial of resource scarcity. People continue to believe that there are more than enough metals available. Yet in 1931, the French writer Paul Valéry warned that 'the time of the finite world has begun',[17] and in 1972 the Club of Rome presented the paradox between the exponential growth of the global population and the economy, and the finiteness of resources.[18] Almost a century has passed since these first warning shots, and our behaviours have still not changed. On the contrary, we consume even more than before. We never dreamed we would catapult ourselves into a world of scarcity the way we did. Evidently, our astounding technological transition outpaced our cognitive progress.

- A lack of mining infrastructure. 'It takes fifteen to twenty-five years to get a mine up and running, from the moment we say "Let's do it" to the time we start extracting minerals,' explained an expert.[19] But according to some projections, a new rare-earths mine will need to be opened every year from now until 2025 to accommodate growth needs.[20] Any delay will cost us dearly in the next two decades. 'We do not produce enough metals today to meet our future needs,' stated an American specialist. 'The numbers just don't add up.'[21]

- Lastly, the energy return on investment (EROI) — the ratio of the energy needed to produce metals to the energy generated using the same metals — is against us. Extracting one to five grams of gold requires crushing one tonne of rocks — up to 10,000 more rocks than

> the metal itself,[22] just like a baker would probably need to mill an entire skip of bread loaves to hope to recover three measly cups of salt. The Italian researcher Ugo Bardi gives another example: 'Imagine that you were asked to take care of the mining waste created by the copper contained inside your new car. An average car contains about 50 kilograms of copper, mainly in the form of wiring. So, on your way home from the dealer, you would be followed by a truck that would then proceed to dump about one tonne of rocks in front of your door.'[23]

How much energy do we need to generate energy? The question may seem hare-brained for most of us, but it's top-of-mind for energy players. One century ago, extracting one hundred barrels of oil required, on average, the energy supplied by one barrel of oil; today this same barrel only produces thirty-five barrels of oil in some drilling areas. Drilling technologies are more efficient, but the most accessible oilfields are now depleted, and more energy is needed to reach new and harder-to-access reserves. For non-conventional crude (shale oil and oil sands), one barrel will produce five barrels at the most. We are teetering on the absurd! Will our production model still be sound when it takes one barrel to fill another?

The same applies to rare metals, which require increasing amounts of energy to be unearthed and refined. Experts state that there are more rare-mineral deposits to be discovered than currently proven, so there is no need to worry about coming up short.[24] But producing these metals takes 7 to 8 per cent of global energy.[25]

What if this ratio were to jump 20 to 30 per cent, or more? Ugo Bardi writes that, in Chile, 'The energy required to mine copper rose by 50 per cent from 2001 and 2010, but the total copper output increased just 13 per cent … The US copper mining industry has also been energy hungry.'[26]

For the same amount of energy, mining companies today extract up to ten times less uranium than they did thirty years ago — and this is true for just about all mining resources. A deposit containing as many minerals as it did in the 1980s would today be considered a 'diamond in the rough' in the mining world.[27] As Bardi concludes, 'The limits to mineral extraction are not limits of quantity; they are limits of *energy*.'[28]

We are starting to see the limits of our production system; they will be reached the day we need more energy to produce the energy we need. But our instinct for conquest compels us to push the envelope to expand humankind's domination in every nook and cranny of the globe (going as far as outer space, as we will discover).

I had to find out more. I took a train to London to peruse ancient maps in the hope that they would provide useful information to assuage our appetite for green growth.

The multiplication of mines

Awaiting me at the Geological Society of London, on the banks of the river Thames, was 'the map that changed the world'.[29] For nearly two centuries, the precious parchment has slumbered in the depths of the GSL archives. To get there, I had to walk through the stately entrance of Burlington House — an imposing building complete with a neo-Renaissance façade overlooking the Piccadilly high street. On the first floor, at the end of a threadbare carpeted staircase, there is a patio with walls lined with old books that serves as a reading room. In the light of two ageing chandeliers, archivist Caroline Lam delicately pieced together fifteen square pages, each one measuring 60 centimetres long and wide. Together they form a treasure map measuring 3 metres by 4 metres — one of the first detailed mining maps in history.

The Great Map is the work of William Smith. Over ten years in the early nineteenth century, the geologist surveyed Great Britain on foot and on horseback with the aim of describing the mineral lie of the land. The copy that is kept at the Geological Society of London is one of the first to have been printed in 1815 — the year in which the map was presented to the public. I needed a magnifying glass to make out the names on the map. Easier to identify were the colour-coded minerals that are plotted in all their diversity: chalk and sand quarries alongside limestone and marble deposits. Drawn in black are the coal seams that would make Great Britain vastly wealthy throughout the nineteenth century.

At the time William Smith published the Great Map, Great Britain was in the throes of its industrial revolution. In the mills, steam generated thermal energy and powered the spinning jennies that would significantly increase productivity. The same steam power led to the introduction of the locomotive on an increasingly dense railway network, which in turn contributed to the phenomenally swift expansion of trade and progress. But to actuate the locomotive pistons and set its wheels in motion, the steam needed to reach a temperature of close to 350 degrees

Celsius. Boilers were therefore fitted with coal furnaces.

This fossil fuel soon became a highly prized resource. 'People needed to know where the deposits were,' said Caroline Lam as she painstakingly put away the fragments of the Great Map. With the help of Smith's map, miners rushed to the coal seams in order to supply the fuel for Great Britain's new energy requirements. In this respect, the Great Map well and truly transformed the world by stimulating the first industrial revolution and giving Great Britain a head start on the rest of Europe. In the Victorian age, the nation used its dominant position in coalmining to establish its industrial, technological, and military superiority, and to become the world's superpower.

Two centuries later, we want to apply the British example to the energy and digital revolution. To secure rare metals supplies, mining maps need to be updated — a realisation brought on by the Chinese embargo that led to states, multinationals, and entrepreneurs scrambling for rare metals. This is being undertaken not only at a national scale, as in Smith's time, but on a planetary scale: deposits of rare earths have been discovered in at least thirty-five countries on five continents. North Korea has some of the most abundant rare earths deposits in the world. In Brazil, President Jair Bolsonaro wants to accelerate the production of niobium — a metal of which Brazil already produces 90 per cent.[30] In the midst of the US–China trade war, Australia is multiplying its mining projects in Western Australia because, in the words of the country's minister of defence, Linda Reynolds, 'It is essential we have a secure source of supply, especially given the current geopolitical headwinds.'[31] Bill Gates, for his part, has even invested in KoBold Metals — a

Californian start-up that promises big data solutions for new cobalt exploration campaigns.[32] Mining companies have already started exploring the hundreds of rare metals deposits around the world.

The trend is anything but sound, with speculative bubbles bursting as mining companies admit that some deposits yield far less than initially announced. Like a spin of the roulette wheel, fortunes are made by the minute while small-time players lose their nest eggs overnight. Either way, the frenzy is creating geopolitical upheaval that goes against the comradely ideals displayed at the conclusion of the Paris accords.

Countries are therefore striking up new alliances for rare metals exploration: Tokyo and Delhi have concluded an export agreement for rare earths mined in India;[33] Japan has deployed its rare-earth diplomacy offensive in Australia, Kazakhstan, and Vietnam; Chancellor Angela Merkel has made numerous trips to Mongolia to sign mining partnerships;[34] South Korean geologists have made official their discussions with Pyongyang on the joint exploration of a deposit in North Korea;[35] France is carrying out prospecting activities in Kazakhstan; Brussels has engaged in economic diplomacy to encourage mining investment with partner states;[36] and in the US, Donald Trump has expressed his interest in buying Greenland — rich in iron, rare earths, and uranium.[37]

This 'diplomatic hodgepodge' of bilateral agreements to secure rare metals supplies signals the end of the bipolar world order inherited from the Cold War, and the muscling in of numerous private and state mining players to the diplomatic arena.

This scramble is upending traditional balances of power. Up until now, Northern Hemisphere countries regularly imposed reprehensible conditions on mineral-rich countries, by and large in the Southern Hemisphere. But the tables are turning: the surge in demand has resulted in a more cautious supply. 'Given the increased competition between consumer countries, it's less a case of the importer deciding to buy the metal than it is of the producer deciding to sell to the buyer,' an expert explained. 'It's called "competitive consumption", and it's a concept we're going to have to come to terms with.'[38]

Several waves of mining nationalism have already put importing states at the mercy of less-powerful supplier countries. Thanks to their mines, the client is no longer (always) king. The geopolitics of rare metals could also give rise to new dominant players, often from the emerging world: Chile, Peru, and Bolivia, thanks to their abundant lithium and copper reserves; India, with its rich titanium, steel, and iron reserves; Guinea and southern Africa, whose subsoils are packed with bauxite, chromium, manganese, and platinum; Brazil, with its abundant bauxite and iron; and New Caledonia, which boasts generous nickel deposits.[39]

The energy and digital transition is sending humanity on a quest for rare metals, and is doomed to aggravate divergence and dissent. Rather than abate the geopolitics of energy, it will compound them.[40] It is a new world that China wants to fashion to its liking, as corroborated by Vivian Wu: 'Given the growth of our domestic demand, we will not be able to meet our own needs within the next five years.' Beijing has therefore begun its own hunt for rare metals, starting in Canada, Australia, Kyrgyzstan, Peru, and Vietnam.[41]

The most prized location of all is Africa, and in particular South Africa, Burundi, Madagascar, and Angola. The former Angolan president José Eduardo dos Santos made rare earths a priority of his mining development strategy to meet Beijing's needs and burnish Angolan–Chinese diplomatic ties.[42] In the Democratic Republic of Congo, China has built a railway line to open up access to the cobalt-rich southern region of Katanga.[43]

The multiplication of mines should spell the end of China's monopoly on rare earths. Is Beijing ready for this sacrifice? Yes and no. The Communist Party wants to have its cake and eat it, too: it wants to alleviate the mining burden while safeguarding its hegemony over the strategic minerals market. And this takes cunning.

From London to Toronto, and Singapore to Johannesburg, not a single conference on rare metals goes by without the same question cropping up, monopolising all discussions: 'What is China playing at?' Indeed, the day after the 2010 embargo, rare-earth prices soared to record highs before taking a nosedive — but for no apparent reason, for supply was just as strong as demand.[44] Many observers believe that Beijing was manipulating prices. 'The Chinese do absolutely whatever they want on the rare-earths market,' deplored Christopher Ecclestone.[45] They can decide to stockpile just as they can decide to slash prices by flooding the market. It has become a headache for non-Chinese mining companies to design long-term economic models with a behemoth like China intentionally destabilising the market. How can they escape bankruptcy when mineral prices are five to ten times lower than forecasted?

The vast majority of alternative projects that emerged after the embargo have been scuppered. The Californian mine Molycorp went bankrupt and reopened, but then had to export its minerals to

China for processing due to a lack of adequate refinery facilities.[46] The Lynas mine in Australia has long been running at a reduced speed, and is being kept afloat by Japan out of its refusal to eat from the hand of its sworn enemy. In Canada, entire battalions of mining companies have shut their doors. Mining licences — once worth their weight in gold — now go for no more than a few hundred dollars.

'The Chinese strategy is not to kill off these projects, but to make them stagnate,' explained Chris Ecclestone. 'Beijing waits, and then makes off with all these mineral deposits for next to nothing.'[47] Once again, while Beijing thinks long-term, Western countries are trapped in short-term logic. The allure of wealth — the catalyst of the mining revival — will not withstand China's chicanery. While the key to sustaining capitalism may very well be rare earths, we would need to mine them in a way that defies logic. The question is whether we can learn from our mistakes.

When China is not undermining the capitalistic foundations of alternative mines, it takes diplomatic action to torpedo them. Such is the case of Kyrgyzstan: the chairman of Stans Energy accused China of putting pressure on the Kyrgyz president to withdraw the Canadian mining house's operating licence without any valid reason.[48] When Beijing doesn't manage to hamper operations, it deploys a strategy of acquiring competing mines. Despite the Chinalco group expressing interest in buying the Mountain Pass mine in California, it was acquired in 2017 by MP Mine Operations LLC — a consortium whose investors include a Chinese mining group, Shenge Resources Shareholding Co. Ltd.[49] China also barges its way into the partial ownership of competing companies:

in Greenland, the same group acquired a sizeable stake in the operations of the Kvanefjeld site, rich in rare earths and uranium. What better way to build up economic intelligence and possibly undermine the emergence of a serious rival?

With its mining-expansion strategy, the Middle Kingdom is working towards a bold objective: to abandon the mining monopolies built on domestic mineral resources in favour of a new dominant position, by controlling the production of a bounty of rare metals across the planet. It's as if Saudi Arabia, which holds the largest proven reserves of oil worldwide, took it upon itself to control the oil reserves of the now thirteen members of OPEC.

China's hegemony over rare metals could continue to grow as the share of renewable energies in our energy mix grows. That is, unless Western countries take a stand and commit to the battle of the mines.

CHAPTER NINE
The last of the backwaters

SHOULD WESTERN COUNTRIES BE REINSTATED AS MINING POWERS? The idea is inconceivable for most, sickening even. In France, environmentalists are against the idea, but President Emmanuel Macron is not.

The idea has been tossed around within the French government for years. It was in one of its glass offices overlooking the Seine that Arnaud Montebourg, the minister of industry under François Hollande, voiced his intention to reopen France's mines: 'The renewal of mining in France is underway ... We want to make sure our country is supplied with the raw materials it needs to ensure its independence, cost and quantity control, and sovereignty.'[1]

France: a slumbering mining giant

Montebourg's words echoed the reindustrialisation policy promised by Hollande during his 2012 presidential campaign. The thinking is sound, for France is in fact a mining giant lying dormant (see Appendix 8). Until the early 1980s, France had a strong mining industry for minerals as diverse as tungsten, manganese, zinc, and antimony, beginning in the first industrial revolution in the nineteenth century. It also had a prosperous iron industry, following the founding of the European Coal and Steel Community (ECSC) in 1952. France's booming mining industry generated thousands of direct and indirect jobs, from the Maurienne valley in the Alps to the rugged landscape of Lorraine in the north-east, and from the Black Mountain folds in the southern centre to the Mouthoumet massif further west. France was on its way to becoming one of the world's biggest antimony, tungsten, and germanium producers.

To turn words into action, Montebourg proposed the creation of a national company of mines, the CMF. It would have a budget of up to €400 million to invest in mining companies, explore partnerships in Africa, and deliver mining permits in mainland France, and thus 'reengage France in the global battle for access to natural resources'.[2]

But following the financial mishaps of two French mining houses, Eramet and Areva, Montebourg's successor, Emmanuel Macron, shelved the project: 'Eramet and Areva were under strain after commodity prices crashed. It would be unwise to create a third fully state-owned enterprise. We prefer to focus on current restructuring.'[3]

The setback didn't stop Macron as president from launching a 'responsible mining' project to reduce the environmental impact of all future mining projects,[4] nor from issuing eleven research permits in metropolitan France and French Guiana.[5] In 2017, the working group, comprising socialists, ecologists, and republicans, presented a mining code reform to parliament to enable sustainable mining in France.[6]

Having the most pro-mining president the country has seen for many years has thus reopened the question of restarting mining in metropolitan France, and fundamentally changed the nature of the debate. Until now, it was easy for France to criticise Beijing for manipulating raw-material prices and ignoring international trade rules. But as France looks to revive its own mining industry, it must face up to its own responsibilities.

The government's position on the controversial topic has polarised public opinion, and sparked citizen movements and resident associations across France, who have taken to the streets, chanting, 'Not here, not anywhere.'[7] One of the environmental groups is the highly vocal Les Amis de la Terre (Friends of the Earth), which has labelled the government's promise of sustainable ore exploration as 'unrealistic', and has denounced the 'mistruths' of the mining revival.[8]

And who can blame them? Numerous mining disasters in France have hardened the population against the industry.[9] The same can be said of many other Western countries. An environmental nonprofit organisation in the US has listed a staggering 500,000 abandoned mines.[10] And according to the Environmental Protection Agency, 'Mining pollutes approximately 40 per cent of the headwaters of Western watersheds and … cleaning up these mines may cost American taxpayers more than $50 billion.'[11] The same has been found in Australia, which has some 60,000 derelict mines that no one knows how to rehabilitate.[12] In the UK, authorities have stated that 'abandoned mines are one of the most significant pollution threats in Britain'.[13] Western citizen movements have gone from NIMBY — Not In My Back Yard — to BANANA — Build Absolutely Nothing Anywhere Near Anything.[14]

But these groups are inconsistent. They condemn the effects of the very world they wish for. They do not admit that the energy and digital transition also means trading oilfields for rare metals deposits, and that the role of mining in the fight against global warming is a reality we have to come to terms with.[15] French government documents state as much: reopening French mines 'is part of the national ecological strategy towards sustainable development'.[16] The debate is also opening our eyes to what the Chinese have known all along: the Western development model is mired in contradictions. It is a hard choice between dreams of a greener world and the reality of a more technological world.

Now that all the makings of an anti-mining rebellion are in place, we shouldn't expect a mining revival in France anytime soon. Even the Ministry of the Environment doesn't foresee any significant mining activity in the next ten years at least. As for the entire rare metals industry, the Government Accountability Office in the US believes it would take at least fifteen years to rebuild the industry.[17] And while Western countries wait for stakeholders to reach an agreement, their mining culture is wasting away. Training is insufficient, and young people are no longer drawn to careers in geology. As the last of the talents disappear, there is a real risk that the sector's revival may be decades in the making.

I support bringing back mining in the West. Not so much for the value, the additional tax revenues, and the thousands of jobs it would create; nor for the strategic security of having our own supply chain at a time when producer countries are tightening the noose around our necks. Rather, my argument is on behalf of the environment.

Reopening mines in the West is the best possible decision we can make for the environment. Relocating our dirty industries has helped keep Western consumers in the dark about the true environmental cost of our lifestyles, while giving other nation-states free rein to extract and process minerals in even worse conditions than would have applied had they still been mined in the West, without the slightest regard for the environment.

The effects of returning mining operations to the West would be positive. We would instantly realise — to our horror — the true cost of our self-declared modern, connected, and green world. We can well imagine how having quarries 'in our backyard' would put an end to our indifference and denial, and drive our efforts to contain the resulting pollution. Because we would not want to live like the Chinese, we would pile pressure onto our governments to ban even the smallest release of cyanide, and to boycott companies operating without the full array of environmental accreditations. We would protest en masse against the disgraceful practice of the planned obsolescence of products, which results in more rare metals having to be mined, and we would demand that billions be spent on research into making rare metals fully recyclable. Perhaps we would also use our buying power more responsibly, and spend more on eco-friendlier mobile phones, for instance. In short, we would be so determined to contain pollution that we would make astounding environmental progress and wind back our rampant consumption.

In this scenario, China's mining activities would face real competition from more ethical Western mines. If China thus lost its edge, its soil, rivers, and air would be beneficiaries. Consumers would be better informed and more demanding, and the industry's only chance at winning back market share would be to improve its own practices. China's environment would emerge the winner from this virtuous competition imposed by the West.

Nothing will change so long as we do not experience, in our own backyards, the full cost of attaining our standard of happiness. Mining the earth responsibly on our own turf will always be more valuable than mining irresponsibly elsewhere. It would be a deeply ecological, selfless, and brave decision that adheres to the ethics of responsibility extolled by countless environmental groups — and rightly so.

An example was the protest in 2009 against European countries exporting their nuclear waste to Russia.[18] Some activists went as far as tying themselves to railway tracks to stop trains coming out of waste-storage facilities, protesting that we alone are responsible for processing our waste, and that palming it off to others is downright immoral. This kind of virulent opposition to the way the end of the fuel cycle is managed should focus just as much on how our dirty mining is outsourced. Protesters should be forming human chains around the ports of Le Havre, Algeciras, or Rotterdam to stop shipments of metals from China from entering the European Customs Union, and should chain themselves to the gates of their government buildings until a law is passed that allows rare earths mining in their countries.

Paris and the conquest of the seas

In France, activists should also ask the government to keep a closer eye on the actions of two Pacific island kings, Filipo Katoa and Eufenio Takala. These enigmatic royals control a wealth of rare earths so great that the future of France's energy and digital transition may depend on their good graces.

On 12 July 2016, wearing the traditional *maros* over their suits, the monarchs from Alo and Sigave — located in the Wallis and Futuna French island collectivity in Far Oceania between Tahiti and New Caledonia — mounted the steps of the Élysée Palace to attend a reception in their honour. While in Paris they were shown around the National Assembly, the Senate, and the Ministry of Overseas Territories. The kings and their delegation had come to discuss matters such as the opening up of their kingdoms, the economic dynamism of their region, and better access to healthcare.[19] It's an important relationship, as shown in 2016 when President Holland invited the kings to his presidential stand alongside the prime minister of New Zealand, John Key, and the US secretary of state, John Kerry, for the Bastille Day parade. Paris is keeping the two royals in its inner circle while it works out how to maintain its presence on their isles. Will it take increasing the €15 million it already spends every year on this special relationship? And will France manage to keep any irredentist hankerings at bay, the consequences of which would be disastrous for the country?

With 16,000 kilometres separating the capital, Mata Utu, from Paris, the Polynesian kingdoms are the furthest French territory from the mainland since becoming French protectorates in 1887.

Yet despite this, nothing in Wallis and Futuna happens according to French standards. Decentralisation laws, international treaties, land regulations, and the traffic code all take second place to local custom, which is guarded by the 'last kings of France' and their mystical powers.[20] Even trade unionists may not strike without first seeking the approval (or ensuring the neutrality) of their 'great chiefdom'.

The kings do not have absolute power, but share it with tribal institutions and with the custodian of antediluvian protocol, the chief of ceremony. The French government maintains the status quo with a monthly salary of €5,500 for the king of Wallis.[21] Indeed, 'Wallis and Futuna are republican kingdoms,' said Pierre Simunek, a former sub-prefect in Mata Utu.[22]

Another infringement of French rule in Wallis and Futuna is the partial application of the 1905 French law on the separation of the church and the state. The basic public service of primary education is entrusted to the church, which by tradition is a powerful institution in Wallis and Futuna. What's more, territorial assemblies are usually opened with a prayer by Bishop Ghislain Marie Raoul Suzanne de Rasilly. 'The sessions start with a blessing from the Holy Spirit, everyone makes the sign of the cross … and then the shouting begins,' recalled Simunek.[23]

The former sub-prefect shared another telling detail. Above the assembly president's seat is a portrait of the French president. Above his portrait are the flags of the Wallis and Futuna kingdoms, which are in turn below the French flag. And topping the entire arrangement is a large crucifix. To Simunek, this implied that Wallis and Futuna's prefect and sub-prefect — the highest representatives of a country known for its aversion to religious symbols — are in this context 'protectors of the throne and altar'.

To what lengths is France prepared to go to maintain its sovereignty over a territory barely bigger than the city of Paris? If France is kowtowing, it is because the isles allow France to stretch its footprint into the Pacific. It gives Paris a say over the world's biggest ocean and largest trade area, and enables it to weigh in on negotiations with regional organisations (such as the Pacific Islands Forum), and to build partnerships with de facto neighbours — and with Australia in particular. As stated in France's white paper on defence: 'New Caledonia and the communities living in French Polynesia and Wallis-et-Futuna make France a political and maritime power in the Pacific.'[24]

The main reason, however, is that Wallis and Futuna give France exclusive access to what in the region is known as La Grande Marmite — a 20-kilometre-diameter crater left over from the Kulo Lasi volcano. It also happens to be a veritable treasure trove of rare earths, and the object of attention from the French Geological Survey, the French Institute for Integrated Marine Science Research, and the Eramet mining group.

Its discovery has created huge tensions in Wallis and Futuna. Fearing that Paris will stake its claim, the chiefdoms are claiming ancestral rights on the territory both above and below water, and Futunans, together with Wallisian elected officials, have demanded the immediate suspension of exploration campaigns. 'One of the ministers of the Royal Counsel even threatened succession over rare earths,' said Simunek.[25]

And so the clear waters of the Wallis and Futuna lagoon are not as placid as they appear to be. But France isn't giving up. 'We need to unite. All of us,' insisted President Hollande in a speech during the state visit of the two Polynesian kings. He concluded his speech on the importance of mining mineral and underwater minerals with a refreshing 'Long live France! Long live Wallis and Futuna! Long live the Republic!'

But they're not the only islands involved: the exclusive economic zones of Tahiti and Clipperton Island, in the north-east Pacific, also have a wealth of rare metals lurking on the ocean floor. Other states have made similar discoveries in the Pacific and Atlantic oceans. Japan recently uncovered staggering quantities of rare earths off the coast of the Ogasawara archipelago, 2,000 kilometres south-east of Tokyo.[26] It would seem that entire swathes of marine areas — 71 per cent of the Earth's surface — are more than watery wastelands where shoals of fish spawn. The 'blue economy' has the potential to generate exponential wealth.

As a new goldrush forms on the horizon, the rare-earths battle (like that of the energy and digital transition) is taking to the seas. Spearheading the offensive is the Canadian group Nautilus. It is preparing to start operations off the coast of Papua New Guinea, and has identified some twenty additional underwater sites to be mined in the future.[27] Always on the ball, China has designed submersibles capable of exploring the ocean floor at record depths. 'Beijing has positioned itself using financing the West doesn't have,' explained a marine geosciences expert. 'The exploration of the oceans has only just begun.'[28] Already, the International Seabed Authority has been flooded with mining-permit applications.

France is also ahead of the pack. It has successfully executed its maritime-extension policy in the last few years by applying the international law of the sea, defined in the Geneva convention of 1958, to encroach on the international maritime areas adjacent to land surfaces under its control, including Guiana, Martinique, Guadeloupe, New Caledonia, and the Kerguelen Islands. France's maritime domain now spans over 11.7 million square kilometres — twenty times mainland France's surface area — making it the world's biggest territory, followed by the US (11.3 million square kilometres) and Australia (8.5 million square kilometres).[29]

Even those nostalgic for its lost colonial empire cannot deny just how big the French republic is today. And it could get even bigger: the Commission on the Limits of the Continental Shelf — the UN body which sets the outer limits of coastal countries — is considering extending the underwater areas owned by France to 350 nautical miles (650 kilometres), provided France can demonstrate that these are natural extensions of its land surface.

France is not the only country to stake its claim on the maritime Monopoly board. Canada, Denmark, Australia, Russia, Japan, Côte d'Ivoire, and Somalia are just some of the dozens of countries requesting the extension of their exclusive economic areas: Denmark has its eyes on the southern continental shelf of Greenland; Russia on parts of the Arctic Ocean; Norway on the Bouvetøya and Dronning Maud Land; Mauritius on the region of Rodrigues island; Papua New Guinea on the Ontong Java Plateau; and the Seychelles on the Northern Plateau Region.[30] Some countries have even resorted to subterfuge: China has gone as far as building artificial islands in the South China Sea so that it can claim exclusive use of the surrounding marine territory.

In short: for thousands of years, 71 per cent of the planet's surface did not belong to anyone; for sixty years, countries owned 40 per cent of the surface of the oceans, and a further 10 per cent is the subject of continental shelf extension requests. We can surmise, therefore, that coastal states have jurisdiction over 57 per cent of the seabed.[31] The allure of rare metals has led to the biggest 'land' grab in history, and in record time.

But there is also something heartening in what we're seeing. For thousands of years, humans the world over have impaled, stabbed, and disembowelled each other over land that makes up one-third of the planet. Now we are deciding on another one-third of the planet — half of the oceans — in no time at all and without having to kill anyone. Only battalions of lawyers armed to the teeth with international law are required. It is significant progress that proves that humanity is capable of improving over time.

The downside, however, is that the exponential growth of our need for rare metals will increasingly commoditise the world's backwaters, which have long been spared from humanity's greed. But it will be decades before mining in the ocean becomes technically and ecologically possible. Until then, states are already divvying up the sea like parcels of land.

The day President Obama fired the starting gun on a new space race

Outer space isn't out of bounds either. This is despite the 1967 Outer Space Treaty, which clearly states that the space beyond the ozone layer is the common property of humanity. But like the oceans, humans are already preparing for minerals from outside the Earth's atmosphere.

The United States was first off the mark. In 2015, President Obama signed off the revolutionary Commercial Space Launch Competitiveness Act, which grants any US citizen the right to 'possess, transport, use, and sell' any space resource. The wording is subtle; it does not openly challenge the international law establishing the principle of non-ownership of celestial bodies, but claims the right of ownership of the riches they bear.[32]

It's a nuance that changes the way we survey the sky. Capitalism — by which objects are ascribed value — has apparently given rise to a new breed of gold-panner who sees asteroids as bags of cash in orbit. Just as a kilogram of apricots goes for a few dollars at the market, an acre of land in Montana or Queensland is worth thousands of dollars, and Modigliani's *Reclining Nude* is auctioned off for US$170 million, an asteroid crossing the Earth's orbit could sell for thousands of billions of dollars.

Since 2015, Americans have started to put price tags on these celestial juggernauts. One of them, the 2011 UW-158, which narrowly missed the Earth's surface in its namesake year, was estimated to be worth €5,000 billion. It was crammed with 90 million tonnes of rare metals — the 'new oil' of the energy and digital transition — including more platinum than humans have ever extracted from the Earth's crust. By opening the door to ownership of space minerals, President Obama gave the self-proclaimed 'space gold-panners' of Silicon Valley the legal comfort they needed to pursue their space mining ambitions. Among these businesses are Space Resources Australia, its US counterpart,

Platinoid Mines Corporation, and its British counterpart, Asteroid Mining Corporation, which has announced the launch of its first asteroid-exploration probe for 2023.[33]

For the time being, commercial space exploitation is utopian at best. The cost of space launches is prohibitive, and there is no ecosystem in place for companies to profit from off-Earth activities. So why didn't the international community simply laugh off the 2015 Space Launch Act? Because space ownership is not a question of how, but of when. So, while member states of the European Space Agency address the taboo concept of space mineral ownership under Obama's Space Launch Act, international agencies will have to provide the legal and diplomatic framework for divvying up the sky.

The UN Office for Outer Space Affairs would also have to review the 1967 treaty to ensure access to ownership if 'New Space' — the fledgling private space economy that has come with the emergence with US space entrepreneurs — is to take off.[34] This could happen within five or six years.[35]

Enter other players like Luxembourg, which is already securing its position.[36] In 2016, its finance minister, Etienne Schneider, announced the Asteroid Mining Plan — the first European space initiative to promote a favourable legal framework for asteroid mining. In 2019, he signed an agreement with the US to share intelligence on space. There is even the provision of €200 million in funding to incentivise space-mining companies to set up in the Grand Duchy. Evidently, the scandal-plagued tax haven is focusing on new growth by fashioning itself as the global hub of the New Space Economy so that it can attract entrepreneurs and jobs ... and generate handsome tax revenues.[37]

We need to think very hard about the moral of this story of land, oceans, and asteroids. The more equal distribution of resources that we celebrate has, in fact, led to the biggest drive for mineral ownership the world has ever seen. The ambition to reduce humans' impact on the ecosystem by way of the energy and digital transition has actually further trampled biodiversity underfoot. And now, our newfound craving for minerals in space is wiping out the last of the sacred wildernesses. Are we looking to the skies to invoke the gods — or subjugate them?

Epilogue

IN THE MID-NINETEENTH CENTURY, WHALE OIL WAS AS ESSENTIAL AS fossil fuels are today. The first industrial revolution in Europe brought the need for better lighting. Lamps using vegetable oils, mineral oils, and animal fat were until then the best way to conquer darkness. Then humans became hooked on whale oil, whose handsome flame would light up households and public streets both effectively and inexpensively. Soon fleets of fortune-seeking whalers were crisscrossing the oceans in search of millions of gallons of the precious fat.

This was the gold rush that created the whaling industry. The sector produced 40 million litres of oil a year, and sparked wars in the Sea of Japan and in the North Pacific over control of prime whaling areas. But so many whales were slaughtered that hunting became increasingly difficult, making oil harder to come by, and hiking the cost of lighting.

Having managed this resource so poorly, would humans learn to do without their precious lighting? Not in the least. In 1853, the

Polish pharmacist Ignacy Łukasiewicz developed a lamp that used a lighter, more functional oil: kerosene. Petroleum would become the next ideal fuel, until electricity become commonplace in the twentieth century.

For many historians and economists, there is a lesson to be learned from our reckless quest for whale oil. Rather than reassess our need for lighting, as our short-sightedness should have taught us, we found a way to illuminate our lives even more by way of petroleum, and from the resilience and prosperity it offered. It's a lesson we need to remember in the twenty-first century as we witness the emergence of many new and abundant energies. Scientists are proposing the implementation of laser and magnetic-confinement fusion, hydrogen-powered and magnetic-levitation vehicles, and even solar power stations placed in the Earth's orbit.[1]

Green technologies will also improve: work is underway to replace the silicon in solar panels with much cleaner and more efficient photovoltaic cells made of a mineral compound called perovskite, and to reduce by two-thirds the carbon dioxide emissions generated by the manufacture of electric batteries.[2] We will also undoubtedly make significant headway in electricity storage, and develop new materials with revolutionary properties. Myriad innovations may render warnings from environmentalists null and void by proving, yet again, that every time an energy source approaches depletion, we have managed to replace it with another.[3] The innovation that saves us from the darkness and confirms the resilience of our species unremittingly wards off what Irish playwright George Bernard Shaw refers to as the 'tragedy' of desire.

But we can draw another lesson from whale oil. The crisis

resulting from its depletion 150 years ago forced us to rethink the way we consume. Yet little came of this introspection, for history repeats itself as new resources run out with every change in the energy model.

And this cycle is unlikely to end anytime soon. Today's new energy technologies will also draw on new raw materials, both natural and synthetic. Polymers, nanomaterials, co-products from industrial processes, bio-based products, and fish waste will become part of our daily lives. We will also turn to hydrogen and thorium, in turn generating their share of environmental waste. Third-generation biofuels sourced from the far reaches of arid deserts and the depths of the oceans will be refined, using highly complex chemical processes. Cooking oil, animal fat, and citrus zest will be collected, using energy-intensive logistical networks. Millions of hectares of forest will be felled and transformed in sawmills of titanic proportions.

The resources of the future will bring new, protean challenges. The question we need to ask ourselves now is: what is the logic behind this next technological leap we all embrace? Can we not see the absurdity of leaping into an environmental sea change that could poison us with heavy metals before we have even seen it through? Can we seriously advocate Confucian harmony through material wellbeing if it means the very opposite: new health problems and environmental chaos?

What is the point of 'progress' if it does not help humanity progress?

Albert Einstein left us with a powerful statement: 'We cannot solve our problems with the same thinking we used when we created them.' Only with a revolution of consciousness can an industrial, technical, and social revolution be meaningful.

This book has sketched out sparse evidence of such leaps of consciousness in the rare metals industry: German manufacturers opting for more expensive tungsten to maintain the diversity of their supply; attempts by Chinese authorities to end the rare-earths black market in Jiangxi province so as to protect the resource; and in Tokyo, Professor Okabe's attempts to recycle metals using salt from the high plains of Bolivia.

For their part, consumers can do more through their own behaviour. The awareness is there, and every one of us already recognise the need to limit our consumption of electronic goods built for obsolescence, to 'eco-design' goods for easy recycling and less waste, to opt for short supply loops, and to focus on saving resources.[4] While moderate consumption does not necessarily lead to 'degrowth', the best energy is that which we use wisely.[5]

I end on this note with French engineer Christian Thomas, who leaves us with a comment of optimism and common sense: 'We don't have a rare material problem; we have a grey matter problem.'[6]

Will we know how to put our grey matter towards finding the antidote to rare metals?

Acknowledgements

IT'S EASY TO THINK THAT WRITING A BOOK IS A LONELY ENDEAVOUR. But it is, in fact, the result of collective work built on contributions, discussions, criticism, and encouragement from numerous sources. It speaks to the interest (often), enthusiasm (sometimes), and kindness (always) of the friends, colleagues, and specialists who honoured me with their involvement. In acknowledging their contribution, which spans several years in some cases, I am also sketching out the landscape of my professional and private life. In particular, I would like to thank:

Hubert Védrine, for writing the preface of this book after reading, annotating, and raising questions, and sometimes contradictions. Our discussions were of inestimable value.

Jean-Paul Tognet, for his honest and sincere accounts, as well as his willingness to read and reread these pages, giving his feedback over the phone from the Ile de Ré.

Christian Thomas, who, from his Paris office adorned in African masks, impressed on me the importance of rigorous analysis.

Paul de Loisy, who, during hours of rich and fascinating discussion on the terrace of a Paris café, helped enhance the book's contents.

Axel Robine, who between flights meticulously and generously read my manuscripts.

Camille Lecomte, who shared her analyses with me, even though we didn't always agree on them!

Didier Julienne, Jack Lifton, and Christopher Ecclestone, who were kind enough to share their considerable expertise, in France, Canada, London, and the United States.

Philippe Degobert, who reviewed the updates on electric motors.

Pierre Simunek, who patiently initiated me in the mysterious ins and outs of Wallisian political life.

Cookie Allez, who helped me mature as a writer.

The Pijac team, and its lifelong chairperson, for their support.

Hélène Crié and Yvan Poisbeau, who provided me with the necessary documents on Rhône-Poulenc.

Randy Henry and the LightHawk team, who gave me access to a light aircraft to fly over the deserts of California and Nevada.

The France-Japan Press Association and Scam, for their financial support of this editorial adventure.

Gérard Tavernier, who set up a number of invaluable meetings.

Félicie Gaudillat, who helped put together a long and comprehensive bibliography.

Stéphanie Berland-Basnier, for her legal advice.

Muriel Steinmeyer, for her loyalty.

Céline Gandner, who initiated the introductions that would set many things in motion.

My sister, Camille Pitron, who supported me in a way only she knows … and many, many more!

Bibliography

Books

BALDÉ, C.P., FORTI V., GRAY, V., KUEHR, R., AND STEGMANN, P., 'The Global E-waste Monitor – 2017, United Nations University (UNU), International Telecommunication Union (ITU) & International Solid Waste Association (ISWA)', Bonn/Geneva/Vienna

BARDI, UGO, *Extracted: how the quest for mineral wealth is plundering the planet*, Chelsea Green Publishing, 2014

BARRÉ, BERTRAND AND BAILLY, ANNE, *Atlas des énergies mondiales: quels choix pour demain?*, Autrement, 3rd edition, 2015

BEFFA, JEAN-LOUIS, *Les Clés de la puissance*, Seuil, 2015

BERGÈRE, MARIE-CLAIRE, *Chine: le nouveau capitalisme d'État*, Fayard, 2013

BIHOUIX, PHILIPPE ET GUILLEBON, BENOÎT (DE), *Quel futur pour les métaux? Raréfaction des métaux: un nouveau défi pour la société*, EDP Sciences, 2010

BIHOUIX, PHILIPPE, *L'Âge des low tech. Vers une civilisation techniquement soutenable*, Seuil, 2014

BILIMOFF, MICHÈLE, *Histoire des plantes qui ont changé le monde*, Albin Michel, 2011

CARTON, MALO AND JAZAERLI, SAMY, *Et la Chine s'est éveillée. La montée en gamme de l'industrie chinoise*, Presses de l'École des mines, 2015

CHALINE, ÉRIC, *Fifty Minerals that Changed the Course of History*, Firefly Books, 2012

CHALMIN, PHILIPPE (DIR.), *Des ressources et des hommes*, Nouvelles Éditions François Bourin, 2016

CHANCEL, CLAUDE AND LE GRIX, LIBIN LIU, *Le Grand Livre de la Chine*, Eyrolles, 2013

COHEN, ÉLIE, *Le Colbertisme high-tech. Économie des télécoms et du grand projet*, Hachette Livre, coll. 'Pluriel', 1992

DEBEIR, JEAN-CLAUDE, DELÉAGE, JEAN-PAUL, AND HÉMERY, DANIEL, *Une histoire de l'énergie*, Flammarion, 2013

DENEAULT, ALAIN AND SACHER, WILLIAM, *Imperial Canada Inc.: Legal Haven of Choice for the World's Mining Industries*, Talonbooks, 2012

DUFOUR, JEAN-FRANÇOIS, *Made by China. Les secrets d'une conquête industrielle*, Dunod, 2012

FLIPO, FABRICE, DOBRÉ, MICHELLE AND MICHOT, MARION, *La Face cachée du numérique. L'impact environnemental des nouvelles technologies*, L'Échappée, 2013

GAIGNERON DE MAROLLES, ALAIN (DE), *L'Ultimatum. Fin d'un monde ou fin du monde?*, Plon, 1984

GIRAUD, PIERRE-NOËL AND OLLIVIER, TIMOTHÉE, *Économie des*

matières premières, La Découverte, coll. 'Repères', 2015

GUILLEBAUD, JEAN-CLAUDE, *Le Commencement d'un monde. Vers une modernité métisse*, Seuil, 2008

HARARI, YUVAL NOAH, *Sapiens: a brief history of humankind*, HarperCollins Publishers, 2015

IZRAELEWICZ, ERIK, *L'Arrogance chinoise*, Grasset, 2011

JUVIN, HERVÉ, *Le mur de l'Ouest n'est pas tombé*, Pierre-Guillaume de Roux, 2015

KAKU, MICHIO, *Physics of the Future: how science will shape human destiny and our daily lives by the year 2100*, Anchor, 2012

KEMPF, HERVÉ, *Fin de l'Occident, naissance du monde*, Seuil, 2013

LAWS, BILL, *Fifty Plants that Changed the Course of History*, Firefly Books, 2010

LE MOIGNE, Rémy, *L'Économie circulaire: comment la mettre en oeuvre dans l'entreprise grâce à la reverse supply chain?*, Dunod, 2014

LENGLET, FRANÇOIS, *La Guerre des empires*, Fayard, 2010

LENGLET, FRANÇOIS, *La Fin de la mondialisation*, Fayard, coll. 'Pluriel', 2014

MEADOWS, DONELLA H., MEADOWS, DENNIS L., RANDERS, JORGEN, and Behrens III, William W., *The Limits to Growth: a report for the Club of Rome's project on the predicament of mankind*, Universe Books, 1972

MOUSSEAU, NORMAND, *Le Défi des ressources minières*, Multi-Mondes Éditions, 2012

'MURKOWSKI, MANCHIN, COLLEAGUES INTRODUCE BIPARTISAN Legislation to Strengthen America's Mineral Security', U.S. Senate Committee on Energy & Natural Resources, 3 May 2019

RABHI, PIERRE, *Vers la sobriété heureuse*, Actes Sud, 2010

RIFKIN, JEREMY, *The Third Industrial Revolution: how lateral power is transforming energy, the economy, and the world*, Palgrave Macmillan, 2011

RIFKIN, JEREMY, *The Zero Marginal Cost Society: the internet of things, the collaborative commons, and the eclipse of capitalism*, Palgrave Macmillan, 2014.

ROGER, ALAIN AND GUÉRY, FRANÇOIS (DIR.), *Maîtres et protecteurs de la nature*, Champ Vallon, 1991

SCHMIDT, ERIC AND COHEN, JARED, *The New Digital Age: reshaping the future of people, nations and business*, Knopf, Random House Inc., 2013.

SURENDRA, M. GUPTA, *Reverse Supply Chains: issues and analysis*, CRC Press, 2013

TINGYANG, ZHAO, *The Tianxia System: an introduction to the philosophy of world institution*, Nanjing, Jiangsu Jiaoyu Chubanshe, 2005

VALÉRY, PAUL, *Regards sur le monde actuel*, Librairie Stock, Delamain et Boutelleau, 193.

WINCHESTER, SIMON, *The Map that Changed the World: William Smith and the birth of modern geology*, HarperCollins, 2001

Essential reading: reports

ARNDT, NICHOLAS (INSTITUTE OF EARTH SCIENCES), AUGÉ, THIERRY (French Geological Survey) and Cuney, Michel (Geo-Resources Laboratoire of the Université de Lorraine), 'Les Ressources minerals en Chine', July 2014

'THE ASIA-PACIFIC MARITIME SECURITY STRATEGY: ACHIEVING US national security objectives in a changing environment', US Department of Defense, 2015

'COMMISSION ON LIMITS OF CONTINENTAL SHELF MEETING AT Headquarters, 11 July–26 August', Background release, United Nations (UN), 11 July 2016

COMMUNICATION FROM THE COMMISSION TO THE EUROPEAN Parliament, the Council, the European Economic and Social Committee, and the Committee of the Regions on the 2017 list of Critical Raw Materials for the EU

'CRITICAL METALS IN THE PATH TOWARDS THE DECARBONISATION OF the EU Energy Sector', Joint Research Centre of the European Commission, 2013

'DEFENSE SCIENCE BOARD TASK FORCE ON HIGH PERFORMANCE Microchip Supply', Office of the Under Secretary of Defense for Acquisition, Technology, and Logistics, 2005

'LES DESSOUS DU RECYCLAGE: DIX ANS DE SUIVI DE LA FILIÈRE DES déchets électriques et électroniques en France', rapport Les Amis de la Terre France, December 2016

DRAFT CRITICAL MINERAL LIST — SUMMARY OF METHODOLOGY AND Background Information — U.S. Geological Survey Technical Input Document in Response to Secretarial Order No. 3359, Open-File Report 2018-1021, U.S. Geological Survey, 2018

'EU SERIOUS AND ORGANISED CRIME THREAT ASSESSMENT (SOCTA)', Europol, 2013

'EXPORT RESTRICTIONS IN RAW MATERIALS TRADE: FACTS, FALLACIES and better practices', OECD, 2014.

EUGSTER, MARTIN AND HISCHIER, ROLAND, 'Key Environmental

Impacts of the Chinese EEE-Industry', Tsinghua University, China, 2007

FREEMAN III, CHARLES W., 'Remember the Magnequench: An Object Lesson in Globalization', *The Washington Quarterly*, 2009, pp. 61–76

'GLOBAL CRUDE STEEL OUTPUT INCREASES BY 4.6 PER CENT IN 2018', World Steel Association, 25 January 2019

GRASSO, VALERIE BAILEY, 'Rare Earth Elements in National Defense: Background, Oversight Issues, and Options for Congress', Congressional Research Service, 23 December 2013

'THE GROWING ROLE OF MINERALS AND METALS FOR A LOW CARBON Future', World Bank Group, June 2017

GUÉHENNO, JEAN-MARIE, French White Paper on Defence and National Security 2013, Ministry of Defence, 2013

H.R.2262 — US COMMERCIAL SPACE LAUNCH COMPETITIVENESS ACT, 114th Congress (2015–2016)

HUISMAN, J., BOTEZATU, I., HERRERAS, L., LIDDANE, M., HINTSA, J., Luda di Cortemiglia, V., Leroy, P., Vermeersch, E., Mohanty, S., van den Brink, S., Ghenciu, B., Kehoe, J., Baldé, C.P., Magalini, F., and Bonzio, A., *Countering WEEE Illegal Trade (CWIT) Summary Report, Market Assessment, Legal Analysis, Crime Analysis and Recommendations Roadmap*, 31 August 2015, Lyon, France

'INTERIOR RELEASES 2018'S FINAL LIST OF 35 MINERALS DEEMED Critical to U.S. National Security and the Economy', *USGS*, May 2018

KORINEK, J. AND KIM, J., 'Export Restrictions on Strategic Raw Materials and their Impact on Trade', OECD Trade Policy

Papers, n° 95, *OECD Publishing*, 2010

'LES ENJEUX STRATÉGIQUES DES TERRES RARES ET DES MATIÈRES premières stratégiques et critiques', report by Patrick Hetzel and Delphine Bataille for the French Parliamentary Office for Science and Technology Assessment (OPECST) Office, n° 617, t. II, 2015–2016, 19 May 2016

LIGER, ALAIN, SECRETARY GENERAL OF THE FRENCH STRATEGIC METALS Committee (COMES), 'Transition énergétique: attention, métaux stratégiques!', The French High Council for Economy, Industry, Energy and Technology, 7 December 2015

MANCHERI, NABEEL, SUNDARESAN, LALITHA, AND CHANDRASHEKAR, S., 'Dominating the World: China and the rare earth industry', National Institute of Advanced Studies, 2013

'MANCHIN, CAPITO & MURKOWSKI REINTRODUCE RARE EARTH ELEMENT Advanced Coal Technologies Act', U.S. Senate Committee on Energy & Natural Resources, 5 April 2019

MARSCHEIDER-WEIDEMANN, FRANK, SABINE, LANGKAU, TORSTEN, Hummen, Lorenz, Erdmann, Tercero, Espinoza Luis, 'Raw Materials for Emerging Technologies 2016', German Mineral Resources Agency (DERA) at the Federal Institute for Geosciences and Natural Resources (BGR), March 2016

'MATIÈRES PREMIÈRES: LE GRAND RETOUR DES STRATÉGIES PUBLIQUES', *Paris Technology Review*, 4 May 2012

MCGREGOR, JAMES, SENIOR COUNSELLOR APCO, 'CHINA'S DRIVE FOR "Indigenous Innovation": a web of industrial policies', US Chamber of Commerce, 2010

MILLS, MARK P., *The Cloud Begins with Coal: big data, big networks, big infrastructure, and big power—an overview of the electricity used by the global digital ecosystem*, August 2013

MINERAL COMMODITY SUMMARIES 2017, U.S. Department of the Interior and U.S. Geological Survey

THE NATIONAL MEDIUM AND LONG-TERM PLAN FOR THE DEVELOPMENT of Science and Technology (2006–2020), The State Council of the People's Republic of China, 2006

O'SULLIVAN, MEGHAN, OVERLAND, INDRA, AND SANDALOW, DAVID, 'The Geopolitics of Renewable Energy', Columbia University Center on Global Energy Policy, June 2017

'PANORAMA DU MARCHÉ DU TUNGSTÈNE', French Geological Survey, July 2012

PARTHEMORE, CHRISTINE AND NAGL, JOHN, 'Fueling the Future Force: preparing the Department of Defense for a post-petroleum era', Center for a New American Security, September 2010

'PRESIDENTIAL EXECUTIVE ORDER ON ASSESSING AND STRENGTHENING the Manufacturing and Defense Industrial Base and Supply Chain Resiliency of the United States', The White House, 21 July 2017

PRESTON, FELIX, BAILEY, ROB AND BRADLEY, SIÂN (Chatham House), 2016, and Jigang, Dr Wei and Changwen, Dr Zhao (DRC), 'Navigating the New Normal: China and global resource governance', January 2016

'PROSPERITY FOR THE MASSES BY 2020 – CHINA'S 13TH FIVE-YEAR PLAN and its business implications', PwC China, Hong Kong and Macau, 2015

'RARE EARTH AND RADIOACTIVE WASTE: A PRELIMINARY WASTE STREAM assessment of the Lynas advanced materials plant, Gebeng, Malaysia', National Toxics Network, April 2012

'RECYCLING RATES OF METALS: A STATUS REPORT', United Nations Environment Programme (UNEP), 2011

'RENEWABLE 2016 GLOBAL STATUS REPORT', Renewable Energy Policy Network for the 21st Century, 2016

'RENEWABLE ENERGY AND JOBS — ANNUAL REVIEW 2018', International Renewable Energy Agency (IRENA)

'REPORT ON CRITICAL RAW MATERIALS FOR THE EU', Report of the Ad Hoc Working Group on Defining Critical Raw Materials, May 2014

'REPORT ON LOW-LEVEL RADIOACTIVE WASTES', Parliamentary Office for Science and Technology Assessment (OPECST), 1992

'RESEARCH AND DEVELOPMENT: USA, EUROPE AND JAPAN INCREASINGLY challenged by emerging countries, says a UNESCO report', *Unescopress*, 10 November 2010

'RESSOURCES MINÉRALES ET ÉNERGIE: RAPPORT DU GROUPE SOL ET sous-sol de l'Alliance Ancre', Alliance nationale de coordination de la recherche scientifique (ANCRE), June 2015

SEAMAN, JOHN, 'Rare Earth and Clean Energy: analyzing China's upper hand', Institut français des relations internationales (IFRI), September 2010

SULLIVAN, J. AND GAINES, L., 'A Review of Battery Life-Cycle Analysis: state of knowledge and critical needs', Argonne National Laboratory, 1 October 2010

'TRENDS IN THE MINING AND METALS INDUSTRY', ICMM (International Council of Mining & Metals), October 2012

'UNESCO SCIENCE REPORT: TOWARDS 2030', 2015

UNITED NATIONS CLIMATE CHANGE, 'Historic Paris Agreement on Climate Change: 195 nations set path to keep temperature rise well below 2 degrees Celsius', 13 December 2015

'UNITED STATES EXPANDS ITS CHALLENGE TO CHINA'S EXPORT Restraints on Key Raw Materials', Office of the United States Trade Representative, July 2016

'U.S. LEADS IN GREENHOUSE GAS REDUCTIONS, BUT SOME STATES ARE Falling Behind', Environmental and Energy Study Institute, 27 March 2018

VAN DER VOET, ESTER, SALMINEN, REIJO, ECKELMAN, MATTHEW, MUDD, Gavin, Norgate, Terry, Hisschier, Roland, Spijker, Job, Vijver, Martina, Selinus, Olle, Posthuma, Leo, de Zwart, Dick, van de Meent, Dik, Reuter, Markus, Tikana, Ladji, Valdivia, Sonia, Wäger, Patrick, Hauschild, Michael Zwicky, and de Koning, Arjan, 'Environmental Risks and Challenges of Anthropogenic Metals Flows and Cycles: a report of the working group on the global metal flows to the International Resource Panel', Kenya, United Nations Environment Programme (UNEP), 2013

'WORLD ENERGY OUTLOOK 2014 FACTSHEET: POWER AND RENEWABLES', International Energy Agency, 2014

Essential reading: articles

APREMONT, BERNARD, 'L'économie de l'URSS dans ses rapports avec la Chine et les démocraties Populaires', *Politique étrangère*, 1956, vol. 21, n° 5, p. 601–13

BUSS, SANDRO, 'Des aimants permanents en terres rares', La Revue polytechnique, no. 1745, 13 April 2010

CHELLANEY, BRAHMA, 'The Challenge from Authoritarian Capitalism to Liberal Democracy', *China-US Focus*, 6 October 2016

'CHINA DECLARED WORLD'S LARGEST PRODUCER OF SCIENTIFIC Articles', *Scientific American*, 23 January 2018

PAARLBERG, ROBERT L., 'Lessons of the Grain Embargo', *Foreign Affairs,* Fall 1980 issue

PETERSEN, JOHN, *'How Large Lithium-ion Batteries Slash EV Benefits'*, 2016

'REP. ALEXANDRIA OCASIO-CORTEZ RELEASES GREEN NEW DEAL Outline', *NPR*, 7 February 2019

VIDAL, OLIVIER, GOFFÉ, BRUNO AND ARNDT, NICHOLAS, 'Metals for a low-carbon society', *Nature Geoscience*, vol. 6, November 2013

Essential viewing

MÖNCH, MAX, LAHL, ALEXANDER, *Ocean's Monopoly*, Werwiewas, 2015

GUILLAUME, PITRON AND TURQUIER, SERGE, *Rare Earths: the dirty war*, 2012

Secrets of the Super Elements, presented by Mark Miodownik, BBC, 2017

TISON, COLINE AND LICHTENSTEIN, LAURENT, *Datacenter: the hidden face of the web*, Camicas Productions, 2012

Useful websites by region

Africa

AFRICAN DEVELOPMENT BANK (ETHIOPIA): https://www.afdb.org

ROYAL BAFOKENG HOLDINGS (SOUTH AFRICA): www.bafokengholdings.com

UNITED NATIONS ENVIRONMENT PROGRAMME (KENYA): www.unep.org

Asia

THE CHINESE SOCIETY OF RARE EARTHS (CHINA): https://fr.linkedin.com/company/the-chinese-society-ofrareearths

NEW ENERGY AND INDUSTRIAL TECHNOLOGY DEVELOPMENT Organization (Japan): www.nedo.go.jp/english

PT TIMAH TBK (INDONESIA): www.timah.com

Australia

ASSOCIATION OF MINING AND EXPLORATION COMPANIES: www.amec.org.au

AUSTRALIA'S MINING MONTHLY: www.miningmonthly.com

AUSTRALIA MINERALS: www.australiaminerals.gov.au

AUSTRALIAN MINING MAGAZINE: www.australianmining.com.au

GEOSCIENCE AUSTRALIA: www.ga.gov.au

LYNAS CORPORATION: www.lynascorp.com

MINERALS COUNCIL OF AUSTRALIA: www.minerals.org.au

NORTHERN MINERALS: https://northernminerals.com.au

France

COMMISSION FOR INDEPENDENT RESEARCH AND INFORMATION about Radiation (CRIIRAD): http://www.criirad.org/english/presentation.html

CYCLOPE: www.cercle-cyclope.com/?lang=en

DIDIER JULIENNE'S ARTICLES ON LE CERCLE, A LES ECHOS NEWSPAPER blog

THE FRENCH ALTERNATIVE ENERGIES AND ATOMIC ENERGY COMMISSION (CEA), Cadarache Centre – New Energy Technologies: www.cea.fr/english

FRENCH GEOLOGICAL SURVEY (BRGM): www.brgm.eu

FRENCH ENVIRONMENT & ENERGY MANAGEMENT AGENCY (ADEME): https://www.ademe.fr/en

GEORESSOURCES GEOLOGY LABORATORY (UNIVERSITÉ DE LORRAINE): https://georessources.univ-lorraine.fr/en

INTERNATIONAL ENERGY AGENCY: www.iea.org

OECD DUE DILIGENCE GUIDANCE FOR RESPONSIBLE SUPPLY CHAINS OF Minerals from Conflict-Affected and High-Risk Areas: http://www.oecd.org/corporate/mne/mining.htm

Europe

BASEL CONVENTION (SWITZERLAND): www.basel.int

EUROPEAN COMMISSION, RAW MATERIAL UNIT (BELGIUM): https://ec.europa.eu/growth/sectors/raw-materials_fr

GLOBAL REPORTING INITIATIVE (NETHERLANDS): www.globalreporting.org

GLOBAL TRADE ALERT (SWITZERLAND): www.globaltradealert.org

WORLD ECONOMIC FORUM (SWITZERLAND): www.weforum.org

INSTITUT FRAUNHOFER (GERMANY): www.fraunhofer.de/en.html

WORLD TRADE ORGANIZATION (SWITZERLAND): www.wto.org

Latin America

GREENPEACE ARGENTINA: www.greenpeace.org/argentina/es

INTERNATIONAL SEABED AUTHORITY (JAMAICA): www.isa.org.jm/fr

Middle East

INTERNATIONAL RENEWABLE ENERGY AGENCY: www.irena.org

United Kingdom

ARGUS MEDIA: www.argusmedia.com

BRITISH GEOLOGICAL SURVEY: www.bgs.ac.uk

CHATHAM HOUSE, THE ROYAL INSTITUTE OF INTERNATIONAL AFFAIRS: www.chathamhouse.org

THE GEOLOGICAL SOCIETY: www.geolsoc.org.uk

INTERNATIONAL TIN RESEARCH INSTITUTE: www.itri.co.uk

THE MINING ASSOCIATION OF THE UNITED KINGDOM: www.mauk.org.uk

WOMEN IN MINING: www.womeninmining.org.uk

United States

AMES LABORATORY: www.ameslab.gov

COMMISSION ON THE LIMITS OF THE CONTINENTAL SHELF (CLCS) OF the Oceans & Law of the Sea (United Nations): www.un.org/depts/los/clcs_new/clcs_home.htm

CONGRESSIONAL RESEARCH REPORTS OF THE FEDERATION OF American Scientists: https://fas.org/sgp/crs

GREAT BASIN RESOURCE WATCH: https://gbrw.org

INVESTORINTEL: https://investorintel.com/sectors/technology-metals

TECHNOLOGY METALS RESEARCH: www.techmetalsresearch.com

U.S. GEOLOGICAL SURVEY: www.usgs.gov

US GOVERNMENT ACCOUNTABILITY OFFICE: www.gao.gov

US MAGNETIC MATERIALS ASSOCIATION: www.usmagneticmaterials.com

WORLD BANK: www.banquemondiale.org

Appendices

Appendix 1

Periodic table of elements

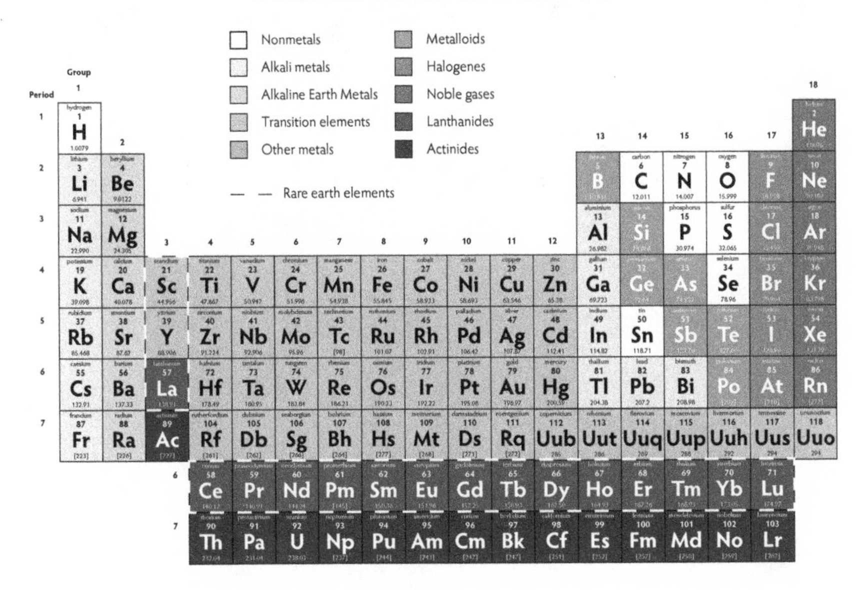

SOURCE: RARE ELEMENT RESOURCES.

Appendix 2

Trends in world primary metal production

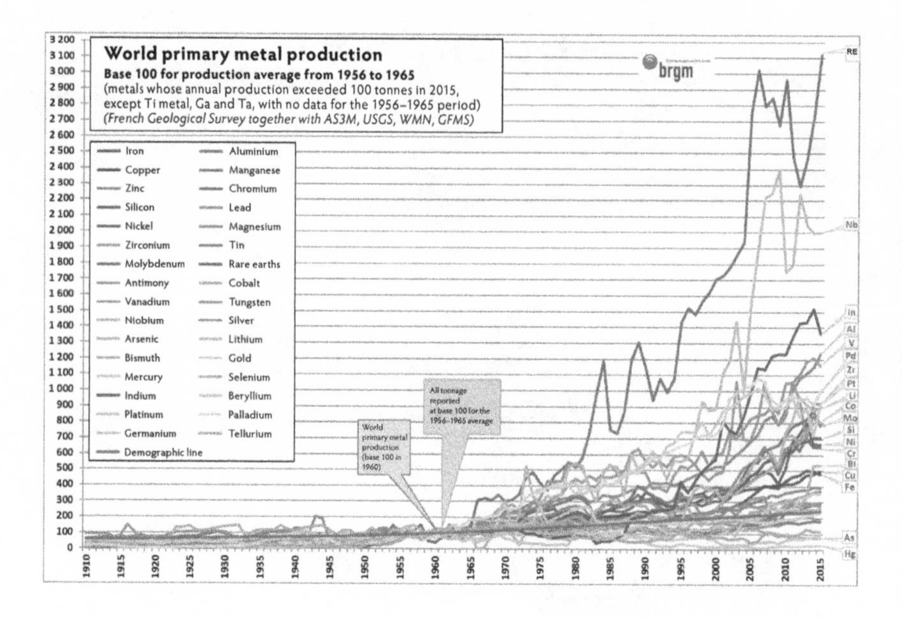

Appendix 3

Countries accounting for the largest share of global supply of critical raw materials

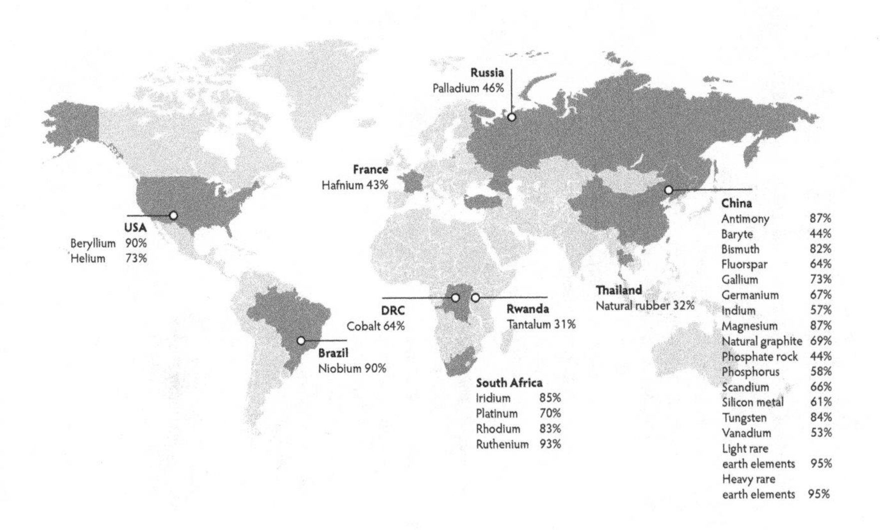

SOURCE: 'STUDY ON THE REVIEW OF THE LIST OF CRITICAL RAW Materials', European Commission, June 2017, https://ec.europa.eu/growth/sectors/raw-materials/specific-interest/critical_en

Appendix 4

China's relative share of global mining and metallurgy production

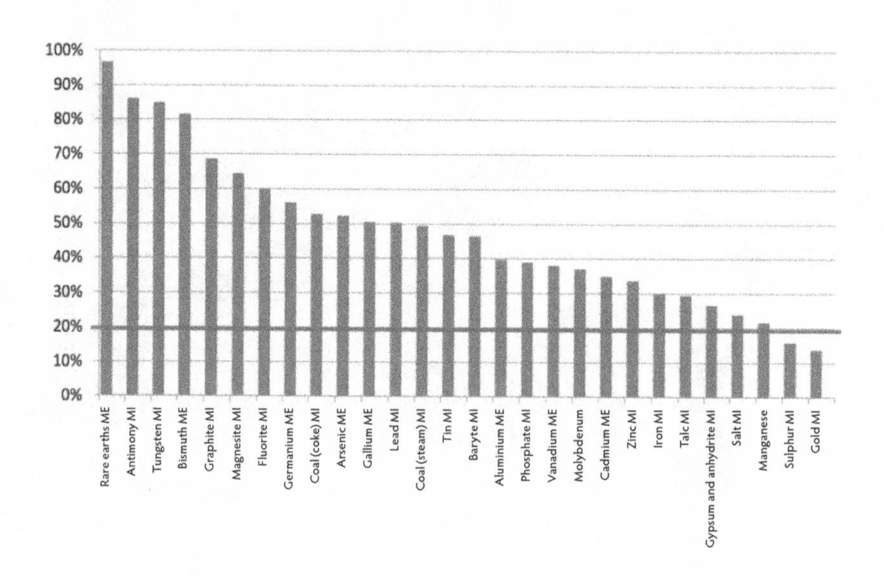

CHINA'S RELATIVE SHARE OF GLOBAL MINING PRODUCTION ('MI') and metallurgy production ('ME') in 2011. The horizontal bar represents the share of the Chinese population in the global population.

SOURCE: WORLD MINING DATA, 2013 EDITION (REF. 63).

Appendix 5

Overview of the rare metals contained in an electric vehicle

SOURCE: 'THE RACE FOR RARE METALS', *The Globe and Mail*, 16 July 2011.

Appendix 6

Rare-metal composition of a smartphone

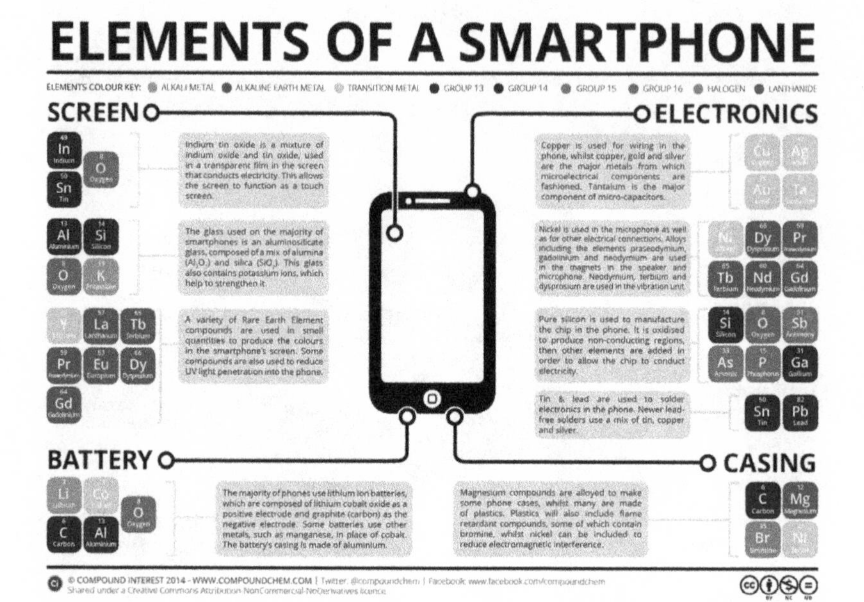

SOURCE: 'THE CHEMICAL ELEMENTS OF A SMARTPHONE', *Compound Interest,* 19 February 2014.

Appendix 7

Summary table of recycling rates of metals

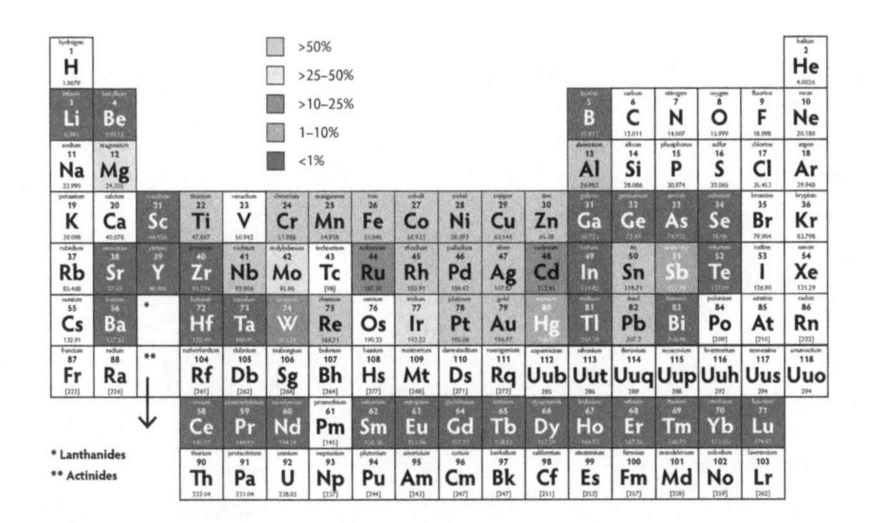

SOURCE: 'RECYCLING RATES OF METALS: A STATUS REPORT', United Nations Environment Programme, 2011.

Appendix 8

France's mining potential

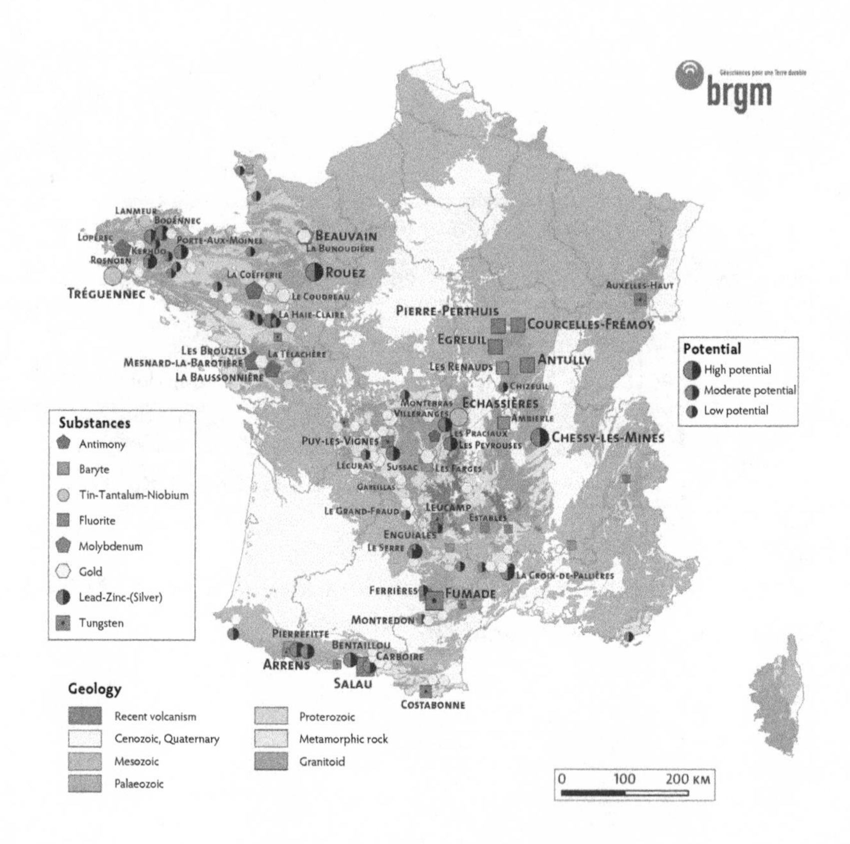

Appendix 9

China's share of world consumption

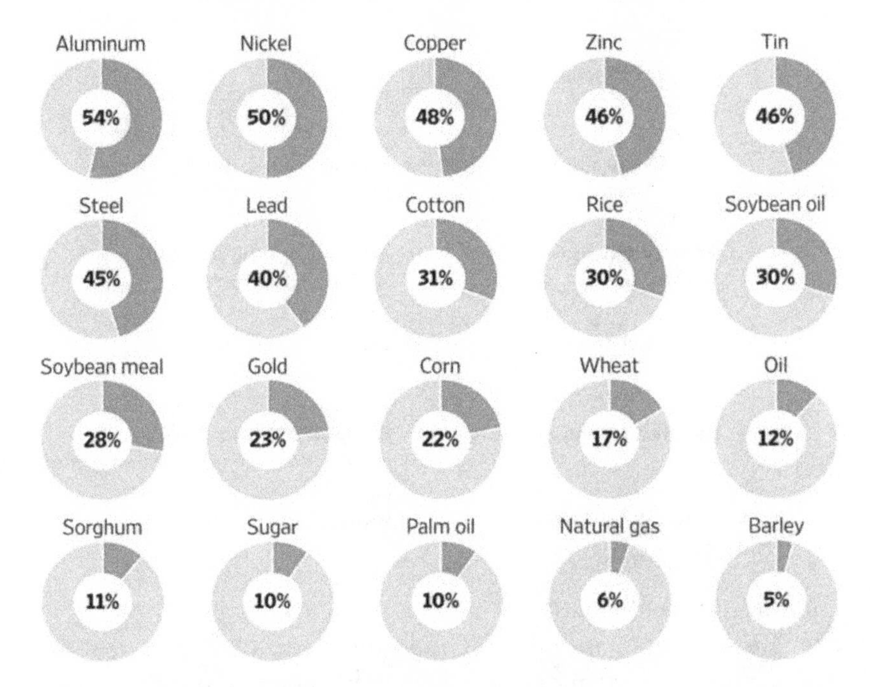

SOURCE: WORLD BUREAU OF METAL STATISTICS (H1 2015 FOR REFINED metals, slab zinc); World Gold Council (2014 for Gold); BP Statistical Review of World Energy 2015 (2014 for oil & natural gas); Metalytics via Morgan Stanley (2015 estimate for finished steel); US Department of Agriculture (2013–14 season for all others). *Wall Street Journal*.

Appendix 10

The life cycle of metals

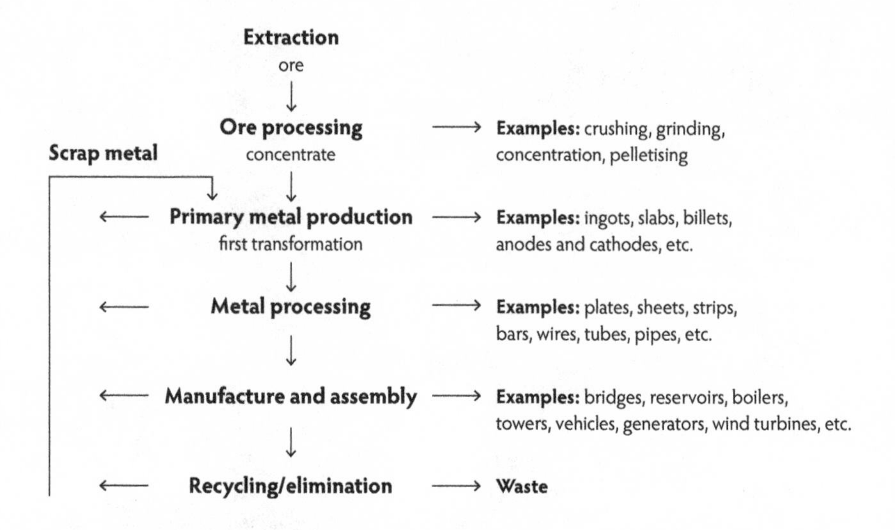

SOURCE: GOUVERNEMENT DU QUÉBEC, MINISTÈRE DE L'ÉNERGIE ET des Ressources naturelles, 'Guide de redaction d'une étude d'opportunité économique et de marché pour la transformation au Québec', The Ministry of Energy and Natural Resources of the Government of Quebec, October 2015, p. 1.

Appendix 11

Main industrial applications of rare minerals

Resource	Applications
Antimony	Fire retardants (additive in plastics), polyethylene terephthalate catalyst
Baryte	Drilling mud for oil and gas drilling, glass industry, radioprotection, healthcare, metallurgy, pyrotechnics
Beryllium	Telecommunications and electronics, aerospace industry, civil and military nuclear power
Bismuth	Thermoelectric generators (automobiles), high-temperature superconductors, lead-free solder
Borate	Glass and ceramics
Cobalt	Mobile phones, computers, hybrid vehicles, magnets
Coking coal	Steel production
Fluorspar	Hydrofluoric acid, steel and aluminium production, ceramics, optics
Gallium	Semi-conductors, light-emitting diodes (LEDs)
Germanium	Photovoltaic cells, fibre optics, catalysis, infrared optics
Indium	Computer chips, LCD screens
Magnesium	Aluminium alloys
Natural graphite	Electric vehicles, aerospace, nuclear industry
Niobium	Satellites, electric vehicles, nuclear industry, jewellery
Silicon metal	Integrated circuits, photovoltaic cells, electric isolators
Tantalum	Miniaturised condenser, superalloys
Tungsten	Cutting tools, shielding, electricity, electronics
Vanadium	Specialty steels, aerospace industry, catalysis

PGMs (platinum-group metals: ruthenium, rhodium, palladium, osmium, iridium, platinum)	Catalysts, jewellery
Rare earths (see table in the following appendix)	Permanent magnets, electric vehicles, wind turbines, TGV (high-speed train), medical scanners, lasers, fibre-optics data transmission, phosphors for plasma screens, security inks for banknotes, catalysis

SOURCE: FRENCH PARLIAMENTARY OFFICE FOR SCIENCE AND Technology Assessment (OPECST), French Geological Survey (BRGM), *Connaissance des énergies*, *Futura-Sciences*, Niobec, Lenntech.

Appendix 12

Main industrial applications of rare earths

Resource	Applications
Lanthanum	Superconductive compounds, lenses, lighting
Cerium	Catalytic converters, oil refinery, metal alloys
Praseodymium	Lighter flint, colourants, magnets
Neodymium	Permanent magnets, autocatalysts, oil refinery, lasers
Promethium	Luminescent compounds
Samarium	Magnets for missiles, permanent magnets, electric motors
Europium	Lasers, nuclear reactors, lighting, geo-chemistry, red phosphors in cathode-ray tubes
Gadolinium	Phosphorescent substance in cathode-ray tubes
Terbium	Green phosphor activator for cathode-ray tubes, permanent magnets
Dysprosium	Permanent magnets, hybrid engines
Holmium	Lasers, magnetism, superconductive compounds
Erbium	Long-distance fibre-optic communication, nuclear medicine
Thulium	Portable radiography, high-temperature superconductors
Ytterbium	Stainless steels, active ion (crystal lasers), portable radiography
Lutetium	Beta emitter (radiation)
Scandium	Lighting, marker, aluminium alloys
Yttrium	Red phosphors in cathode-ray tubes, superconductor alloys, fire bricks, fuel cells, magnets

SOURCE: FRENCH SENATE, BRITISH GEOLOGICAL SURVEY, ECONOMIC Warfare School (EGE), Congressional Research Service, Portail de l'Intelligence Economique.

Appendix 13

The European Commission's list of critical raw materials for the EU

Raw materials	Main global producers (average 2010–2014)	Main importers to the EU (average 2010–2014)	Sources of EU supply (average 2010–2014)	Import reliance rate*	Substitution indexes EI/ SR**	End-of-life recycling input rate***
Antimony	China (87%) Vietnam (11%)	China (90%) Vietnam (4%)	China (90%) Vietnam (4%)	100%	0.91 / 0.93	28%
Baryte	China (44%) India (18%) Morocco (10%)	China (53%) Morocco (37%) Turkey (7%)	China (34%) Morocco (30%) Germany (8%) Turkey (6%) United Kingdom (5%) Other EU (4%)	80%	0.93 / 0.94	1%
Beryllium	United States (90%) China (8%)	n/a	n/a	n/a****	0.99 / 0.99	0%

(*) The 'Import reliance rate' takes into account global supply and actual EU sourcing in the calculation of Supply Risk, and it is calculated as follows: EU net imports / (EU net imports + EU domestic production).

(**) The 'Substitution index' is a measure of the difficulty in substituting the material, scored and weighted across all applications, calculated separately for both Economic Importance and Supply Risk parameters. Values are between 0 and 1, with 1 being the least substitutable. The economic importance is corrected by the Substitution Index (SIEI) related to technical and cost performance of the substitutes for individual applications of each material. The supply risk is corrected by the Substitution Index (SISR) related to global production, criticality and co-/by-production of the substitutes for individual applications of each material.

(***) The 'End-of-life recycling input rate' measures the ratio of recycling from old scrap to EU demand of a given raw material, the latter equal to primary and secondary material supply inputs to the EU.

(****) The EU import reliance cannot be calculated for beryllium, as there is no production and trade for beryllium ores and concentrates in the EU.

Raw materials	Main global producers (average 2010–2014)	Main importers to the EU (average 2010–2014)	Sources of EU supply (average 2010–2014)	Import reliance rate	Substitution indexes EI/SR	End-of-life recycling input rate
Bismuth	China (82%) Mexico (11%) Japan (7%)	China (84%)	China (84%)	100%	0.96 / 0.94	1%
Borate	Turkey (38%) United States (23%) Argentina (12%)	Turkey (98%)	Turkey (98%)	100%	1.0 / 1.0	0%
Cobalt	Democratic Republic of Congo (64%) China (5%) Canada (5%)	Russia (91%) Democratic Republic of Congo (7%)	Finland (66%) Russia (31%)	32%	1.0 / 1.0	0%
Coking coal	China (54%) Australia (15%) United States (7%) Russia (7%)	United States (39%) Australia (36%) Russia (9%) Canada (8%)	United States (38%) Australia (34%) Russia (9%) Canada (7%) Poland (1%) Germany (1%) Czech Republic (1%) United Kingdom (1%)	63%	0.92 / 0.92	0%
Fluorspar	China (64%) Mexico (16%) Mongolia (5%)	Mexico (38%) China (17%) South Africa (15%) Namibia (12%) Kenya (9%)	Mexico (27%) Spain (13%) China (12%) South Africa (11%) Namibia (9%) Kenya (7%) Germany (5%) Bulgaria (4%) United Kingdom (4%) Other EU (1%)	70%	0.98 / 0.97	1%
Gallium*	China (85%) Germany (7%) Kazakhstan (5%)	China (53%) United States (11%) Ukraine (9%) South Korea (8%)	China (36%) Germany (27%) United States (8%) Ukraine (6%) South Korea (5%) Hungary (5%)	34%	0.95 / 0.96	0%
Germanium	China (67%) Finland (11%) Canada (9%) United States (9%)	China (60%) Russia (17%) United States (16%)	China (43%) Finland (28%) Russia (12%) United States (12%)	64%	1.0 / 1.0	2%
Hafnium	France (43%) United States (41%) Ukraine (8%) Russia (8%)	Canada (67%) China (33%)	France (71%) Canada (19%) China (10%)	9%	0.93 / 0.97	1%

* Gallium is a by-product; the best available data refer to production capacity, not to production as such.

Raw materials	Main global producers (average 2010–2014)	Main importers to the EU (average 2010–2014)	Sources of EU supply (average 2010–2014)	Import reliance rate	Substitution indexes EI/ SR	End-of-life recycling input rate
Helium	United States (73%) Qatar (12%) Algeria (10%)	United States (53%) Algeria (29%) Qatar (8%) Russia (8%)	United States (51%) Algeria (29%) Qatar (8%) Russia (7%) Poland (3%)	96%	0.94 / 0.96	1%
Indium	China (57%) South Korea (15%) Japan (10%)	China (41%) Kazakhstan (19%) South Korea (11%) Hong Kong (8%)	China (28%) Belgium (19%) Kazakhstan (13%) France (11%) South Korea (8%) Hong Kong (6%)	0%	0.94 / 0.97	0%
Magnesium	China (87%) United States (5%)	China (94%)	China (94%)	100%	0.91 / 0.91	9%
Natural graphite	China (69%) India (12%) Brazil (8%)	China (63%) Brazil (13%) Norway (7%)	China (63%) Brazil (13%) Norway (7%) EU (< 1%)	99%	0.95 / 0.97	3%
Natural rubber	Thailand (32%) Indonesia (26%) Vietnam (8%) India (8%)	Indonesia (32%) Malaysia (20%) Thailand (17%) Ivory Coast (12%)	Indonesia (32%) Malaysia (20%) Thailand (17%) Ivory Coast (12%)	100%	0.92 / 0.92	1%
Niobium	Brazil (90%) Canada (10%)	Brazil (71%) Canada (13%)	Brazil (71%) Canada (13%)	100%	0.91 / 0.94	0.3%
Phosphate rock	China (44%) Morocco (13%) United States (13%)	Morocco (31%) Russia (18%) Syria (12%) Algeria (12%)	Morocco (28%) Russia (16%) Syria (11%) Algeria (10%) EU – Finland (12%)	88%	1.0 / 1.0	17%
Phosphorus	China (58%) Vietnam (19%) Kazakhstan (13%) United States (11%)	Kazakhstan (77%) China (14%) Vietnam (8%)	Kazakhstan (77%) China (14%) Vietnam (8%)	100%	0.91 / 0.91	0%
Scandium	China (66%) Russia (26%) Ukraine (7%)	Russia (67%) Kazakhstan (33%)	Russia (67%) Kazakhstan (33%)	100%	0.91 / 0.95	0%
Silicon metal	China (61%) Brazil (9%) Norway (7%) United States (6%) France (5%)	Norway (35%) Brazil (18%) China (18%)	Norway (23%) France (19%) Brazil (12%) China (12%) Spain (9%) Germany (5%)	64%	0.99 / 0.99	0%

Raw materials	Main global producers (average 2010–2014)	Main importers to the EU (average 2010–2014)	Sources of EU supply (average 2010–2014)	Import reliance rate	Substitution indexes EI/ SR	End-of-life recycling input rate
Tantalum*	Rwanda (31%) Democratic Republic of Congo (19%) Brazil (14%)	Nigeria (81%) Rwanda (14%) China (5%)	Nigeria (81%) Rwanda (14%) China (5%)	100%	0.94 / 0.95	1%
Tungsten**	China (84%) Russia (4%)	Russia (84%) Bolivia (5%) Vietnam (5%)	Russia (50%) Portugal (17%) Spain (15%) Austria (8%)	44%	0.94 / 0.97	42%
Vanadium	China (53%) South Africa (25%) Russia (20%)	Russia (71%) China (13%) South Africa (13%)	Russia (60%) China (11%) South Africa (10%) Belgium (9%) United Kingdom (3%) Netherlands (2%) Germany (2%) Other EU (0.5%)	84%	0.91 / 0.94	44%
Platinum Group Metals	South Africa (83%) iridium, platinum, rhodium, ruthenium Russia (46%) palladium	Switzerland (34%) South Africa (31%) United States (21%) Russia (8%)	Switzerland (34%) South Africa (31%) United States (21%) Russia (8%)	99.6%	0.93 / 0.98	14%
Heavy Rare Earth Elements	China (95%)	China (40%) USA (34%) Russia (25%)	China (40%) USA (34%) Russia (25%)	100%	0.96 / 0.89	8%
Light Rare Earth Elements	China (95%)	China (40%) USA (34%) Russia (25%)	China (40%) USA (34%) Russia (25%)	100%	0.90 / 0.93	3%

(*) Tantalum is covered by the Conflict Minerals Regulation (Regulation (EU) 2017/821) establishing a Union system for supply chain due diligence to curtail opportunities for armed groups and security forces to trade in tin, tantalum and tungsten, and their ores, and gold.

(**) Tungsten is covered by the Conflict Minerals Regulation (Regulation (EU) 2017/821) establishing a Union system for supply chain due diligence to curtail opportunities for armed groups and security forces to trade in tin, tantalum and tungsten, and their ores, and gold.

SOURCE: COMPILED ON THE BASIS OF THE FINAL REPORT OF THE 'STUDY ON THE REVIEW of the list of Critical Raw Materials' 2017.

Appendix 14

Lifespan of the viable reserves of the principal metals needed for the energy transition

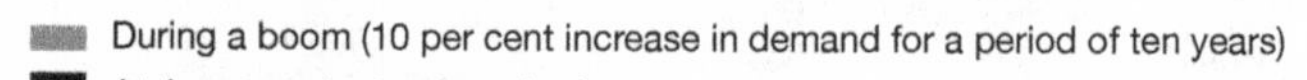

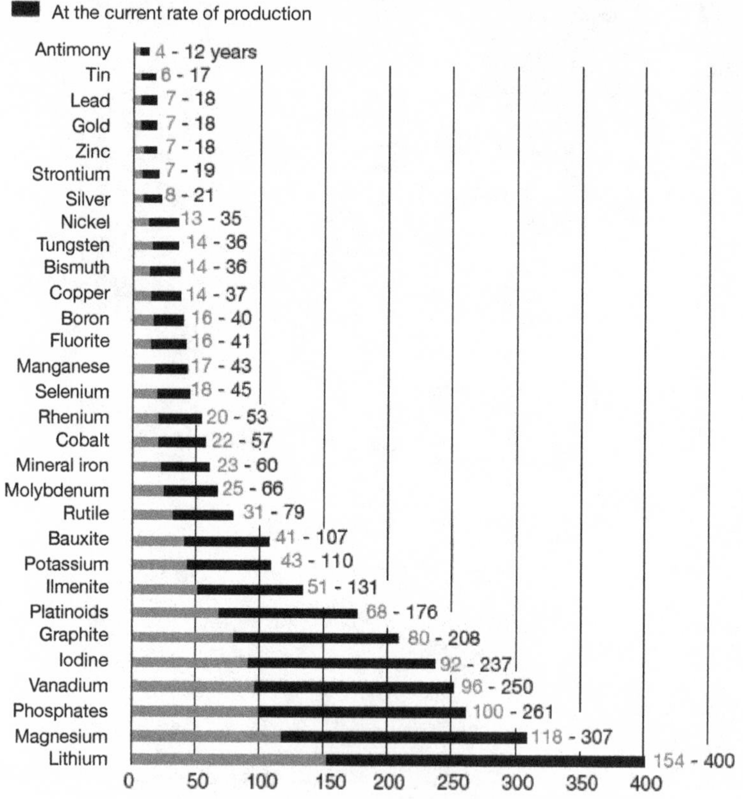

SOURCE: TABLE BY L. PENNEC FOR *L'Usine Nouvelle*, 2017.

Notes

Introduction

1 Yuval Noah Harari, *Sapiens: a brief history of humankind*, HarperCollins Publishers, 2015.
2 United Nations Climate Change, 'Historic Paris Agreement on Climate Change: 195 Nations Set Path to Keep Temperature Rise Well Below 2 Degrees Celsius', 13 December 2015.
3 Distilling a tonne of orange blossom petals only produces a kilogram of essential oil.
4 It takes 500 kilograms of coca leaves to produce a kilogram of cocaine.
5 On average, a kilogram of rock contains 120 milligrams of vanadium, 66.5 milligrams of cerium, 19 milligrams of gallium, and 0.8 milligrams of lutecium.
6 He and his colleagues, Jacob A. Marinsky and Lawrence E. Glendenin, produced it at the Oak Ridge National Laboratory in 1945.
7 Jeremy Rifkin, *The Third Industrial Revolution: how lateral power is transforming energy, the economy, and the world*, Palgrave Macmillan, 2011.
8 Since 2013, the Hauts-de-France region has availed itself of Jeremy Rifkin's advisory services to develop new energy-consumption models based on the crossover of green and digital technologies. See www.businessinsider.fr/us/jeremy-rifkin-interview-2017-6.
9 Renewable energy includes other types of energy, such as hydraulic energy, biofuels, and biomass. Read the 'Renewable 2016 Global Status Report', Renewable Energy Policy Network for the 21st Century, 2016.
10 Christine Parthemore and John Nagl, 'Fueling the Future Force: preparing the Department of Defense for a post-petroleum era', Center

for a New American Security, September 2010. Also read the article 'U.S. Military Marches Forward on Green Energy, despite Trump', *Reuters*, 1 March 2017. According to the author, 'the number of [US] military renewable energy projects nearly tripled to 1,390 between 2011 and 2015 ... Many of those projects are at U.S. bases, where renewable energy allows the military to maintain its own independent source of power in case of a natural disaster or an attack — or cyber-attack — that disables the public grid.'

11 Instead, armies would rely on small renewable-energy plants that are less vulnerable to enemy attacks. Read Ugo Bardi, *Extracted: how the quest for mineral wealth is plundering the planet*, Chelsea Green Publishing, 2014.

12 See Hervé Juvin, *Le mur de l'Ouest n'est pas tombé* [*The Western Wall Did Not Fall*], Pierre-Guillaume de Roux, 2015.

13 'BP Says World's Oil Consumption Will Peak in Late 2030s', BBC News, 21 February 2018.

14 The renewable-energy economy will create 28.8 million jobs by 2050 in leading industries worldwide, according to 'Renewable Energy and Jobs — Annual Review 2018', International Renewable Energy Agency (IRENA), 2017.

15 See the 'white paper' by Florentin Krause, Hartmut Bossel, and Karl-Friedrich Müller-Reißmann, *Energie-Wende: Wachstum und Wohlstand ohne Erdöl und Uran* [*Energy Transition: growth and wellbeing without crude oil and uranium*], S. Fischer Verlag, 1980.

16 The 196 delegations comprised 195 states and the European Union.

17 At no point does the Paris agreement on climate change include the words 'metals', 'minerals', or 'commodities'. Similarly, none of the decisions made at the COP 24 in Katowice (Poland) in December 2018 addresses mineral resources. As stated by the press service of the United Nations Framework Convention on Climate Change, 'we are not aware of a specific discussion on the question of mineral resources'.

18 Most rare earths cannot be substituted. Refer to the communication from the Commission to the European Parliament, the Council, the European Economic and Social Committee, and the Committee of the Regions on the 2017 list of Critical Raw Materials for the EU, 13 September 2017, p. 4 and following. Also refer to the substitution indexes EI/SR in the list of critical raw materials for the EU under Annex 1 of the communication.

19 'US Admiral Warns: only war can now stop Beijing controlling the South China Sea', *News.com.au*, 22 April 2018.

20 '2019 Revision of World Population Prospects', Department of Economic and Social Affairs Population Division, United Nations, New York, 2019.

21 Bill Laws, *Fifty Plants that Changed the Course of History*, Firefly Books, 2010.

Chapter One: The rare metals curse

1 Gallium, for example, is a by-product of aluminium. Selenium and tellurium are associated with copper. Indium and germanium are zinc by-products. See Philippe Bihouix and Benoît de Guillebon, *Quel futur pour les métaux? Raréfaction des métaux: un nouveau défi pour la société* [*What Does the Future Hold for Metals? Rarefication of Metals: a new challenge for society*], EDP Sciences, 2010, p. 33.

2 Updated in May 2018, the USGS list includes thirty-five mineral commodities such as caesium, chrome, lithium, rubidium, uranium, and strontium. The inventory prepared by the European Commission includes twenty-seven mineral commodities: antimony, baryte, beryllium, bismuth, borate, cobalt, coking coal, fluorspar, gallium, germanium, hafnium, helium, indium, magnesium, natural graphite, natural rubber, niobium, phosphate rock, phosphorus, scandium, silicon metal, tantalum, tungsten, vanadium, platinum group metals, heavy rare-earth elements, and light rare-earth elements. Some of the metals listed, such as silicon, are not considered rare by geologists. The European Commission nevertheless qualifies them as 'critical' due to the threat to their supply. Shortages can often result from a lack in mining and refining infrastructure, adding the notion of industrial rarity to that of geological rarity. The scientific community uses the term 'rare metals' in reference to these two criteria. Refer to Appendix 13 for the list of metals defined as 'rare' by the European Commission.

3 Refer to the criteria set out in Report 782, *Key Issues with Strategic Metals: the case of rare earths* (submitted 23 August 2011) by Claude Birraux and Christian Kert, deputies of the French Parliamentary Office for Science and Technology Assessment (OPECST). For a US critical mineral classification methodology, refer to the Draft Critical Mineral List — Summary of Methodology and Background Information — U.S. Geological Survey Technical Input Document in Response to Secretarial Order No. 3359, Open-File Report 2018-1021, U.S. Department of the Interior, U.S. Geological Survey.

4 Harari, *Sapiens*, op. cit.

5 Such as praseodymium and neodymium.

6 'Travel to the Largest and Most Powerful MRI Magnet in the World', The French Alternative Energies and Atomic Energy Commission (CEA), 2 May 2017.

7 These super magnets are produced with the rare-earth minerals neodymium and samarium that are alloyed with other metals, such as iron, boron, and cobalt. Magnets are usually 30 per cent neodymium and 35 per cent samarium. For the sake of clarity, they are more commonly referred to as 'rare-earth magnets' by the scientific community.

8 The vehicles of car manufacturers Toyota, Nissan, Mitsubishi, General

Motors, PSA, and BMW contain rare-earth magnets. But others have done without, such as US manufacturer Tesla's squirrel cage induction motorcar and Renault's rotor-coil engine Zoé. Nevertheless, the two are bigger and heavier than engines with rare-earth magnets. Interview with Philippe Degobert, electrical engineering lecturer at the École nationale supérieure des Arts et des Métiers (ENSAM) and director of Masters in Mobility and Electric Vehicles (MVE), 2017.

9 In 2016, seven of the ten most powerful wind turbines were made using rare-earth metals (the V164 by Vestas, the AD-180 and ADS-135 by Adwen, the SWT 8.0 by Siemens, the 6 MW Haliade by General Electric, the SCD 6.0 by Ming Yang, and the Dong Fang/Hyundai 5.5 MW). German wind-turbine manufacturer Enercon opted for separately excited annular generation, as it claims that it is possible to do without permanent magnets. The magnetic fields needed to generate electricity are created electrically. Interview with Philippe Degobert, electrical engineering lecturer at École nationale supérieure des Arts et des Métiers (ENSAM) and director of Masters in Mobility and Electric Vehicles (MVE), 2017.

10 Especially solar panels made from copper indium gallium selenide (CIGS) solar cells.

11 According to John Ormerod, an expert in magnets and founder of the consultancy JOC LLC, most electric engines in operation in the world today work by induction and therefore do not use rare-earth magnets. These engines, used mostly in heating, ventilation, and air-conditioning devices, are inexpensive but also low-performance. But industrialising high-performance engines will mean having to use rare-earth magnets. This is the case for the engines used in electric vehicles and certain wind turbines. Higher demand for electric cars will in future lead to more rare-earth magnets being used in engines. Interview with John Ormerod, JOC LLC, 2017.

12 Light-emitting diode (LED) bulbs.

13 'Renewable Energy and Jobs — Annual Review 2018', International Renewable Energy Agency (IRENA).

14 '100 Per Cent Clean and Renewable Wind, Water, and Sunlight (WWS) All-Sector Energy Roadmaps for the 50 United States', Royal Society of Chemistry, 27 May 2015.

15 'Rep. Alexandria Ocasio-Cortez Releases Green New Deal Outline', *NPR*, 7 February 2019.

16 Gold, copper, lead, silver, tin, mercury, and iron.

17 British Petroleum 2017: Outlook for 2035. 'Gas is the fastest growing fuel (1.6 per cent p.a.); Oil continues to grow (0.7 per cent p.a.), although its pace of growth is expected to slow gradually; The growth of coal is projected to decline sharply: 0.2 per cent p.a. compared with 2.7 per cent p.a. over the past 20 years — coal consumption is expected to peak in the mid-2020s; Renewable energy is the fastest growing source

of energy (7.1 per cent p.a.), with its share in primary energy increasing to 10 per cent by 2035, up from 3 per cent in 2015' (p. 15).

18 '*Quand le monde manquera de métaux*' ['When the World Runs Out of Metals'], *Basta Mag*, 26 September 2012.

19 Read Frank Marscheider-Weidemann, Sabine Langkau, Torsten Hummen, Lorenz Erdmann, and Luis Tercero Espinoza, 'Raw Materials for Emerging Technologies 2016', German Mineral Resources Agency (DERA), Federal Institute for Geosciences and Natural Resources (BGR), March 2016.

20 A rough estimate in the high range.

21 'Environmental Disaster Strains China's Social Fabric', *The Financial Times*, 26 April 2006.

22 'Toxic Mine Spill Was Only Latest in Long History of Chinese Pollution', *The Guardian*, 14 April 2011.

23 'Dwindling Supplies of Rare Earth Metals Hinder China's Shift from Coal', *TrendinTech*, 7 September 2016.

24 The first opium war pitched China against the United Kingdom from 1839 to 1842, and was led by France, the United Kingdom, Russia, and the United States. The second opium war would last from 1856 to 1860.

25 The German concessions in Shandong, a province in the north of China, were handed over to Japan.

26 Founded by Sun Yat-sen, the Kuomintang party was defeated by the communist regime in 1949.

27 Read the speech of Patrice Christmann from the French Geological Survey (BRGM) in the minutes of the public hearing of 6 July 2015 on the implementation of the French Parliamentary Office for Science and Technology Assessment (OPECST) policy for rare-earth metals and strategic and critical raw materials.

28 Interview with Thomas Kruemmer, managing director of Kloeckner Metals, 2016.

29 Interview with consultant Bruno Gensburger, Mutandis consultancy, 2016.

30 'China Cuts Smog but Health Damage Already Done: study', *Reuters*, 17 April 2018.

31 'The Cobalt Pipeline: tracing the path from deadly hand-dug mines in Congo to consumers' phones and laptops', *The Washington Post*, 30 September 2016.

32 See '*Le chrome (Cr) — éléments de criticité*' ['Chrome (Cr) — elements of criticality'], French Geological Survey (BRGM), July 2017. The European Commission does not consider chrome a critical metal, unlike the US, which included it in its list, updated in 2018. See Interior Releases 2018's 'Final List of 35 Minerals Deemed Critical to U.S. National Security and the Economy', *U.S. Geological Survey*, 18 May 2018.

33 'Kazakh Ecologists: Syr Darya waters poisonous', *Ferghana News*

Agency, 9 April 2015.

34 'Lithium Squeeze Looms as Top Miner Front-loads, Chile Says', *Mining Weekly*, 26 June 2017.

35 'Chile's Supreme Court Casts Shadow over Barrick's Plans to Restart Pascua-Lama', *Mining.com*, 15 March 2017.

36 'In Argentina's Salar de Hombre Muerto, Local Communities Claim that Lithium Operations Have Contaminated Streams Used for Humans, Livestock and Crop Irrigation', *Friends of the Earth*, 2013.

37 'The World's Worst Pollution Problems 2016: the toxics beneath our feet', *Green Cross Switzerland and Pure Earth*, 2016.

Chapter Two: The dark side of green and digital technologies

1 The 5th Annual Cleantech & Technology Metals Summit: Invest in the Cleantech Revolution.

2 'World Energy Outlook 2019 Factsheet: Power and renewables', International Energy Agency, 2019.

3 Tourillon reached this estimate using the Greenhouse Gas Equivalencies Calculator of the United States Environmental Protection Agency. Refer to the EPA's Greenhouse Gas Equivalencies Calculator online.

4 The sun's rays generate thermal energy, which heats fluids such as water that can then be used directly (using a solar water heater) or indirectly (steam passing through a generator to produce electricity).

5 Julia Bucknall, 'Cutting Water Consumption in Concentrated Solar Power Plants', *The Water Blog*, 20 May 2013.

6 Kimberly Aguirre, Luke Eisenhardt, Christian Lim, Brittany Nelson, Alex Norring, Peter Slowik, and Nancy Tu, 'Lifecycle Analysis Comparison of a Battery Electric Vehicle and a Conventional Gasoline Vehicle', UCLA Institute of the Environment and Sustainability, June 2012. For more information on the environmental impacts of electric batteries, see J. Sullivan and L. Gaines, 'A Review of Battery Life-cycle Analysis: state of knowledge and critical needs', Argonne National Laboratory, 1 October 2010.

7 Read 'Extraordinary Raw Materials in a Tesla Model S', *Visual Capitalist*, 7 March 2016.

8 'The 100 kWh battery also increases range substantially to an estimated 315 miles [507 kilometres] on the EPA cycle and 613 km on the NEDC cycle, making it the first to go beyond 300 miles [482 kilometres] and the longest range production electric vehicle by far'; 'New Tesla Model S Now the Quickest Production Car in the World', *Tesla*, 23 August 2016.

9 'Musk: Millions of Teslas, 500-mile range coming', *CNBC*, 6 November 2015. Musk may have stepped down as one of the CEOs

of President Trump's business advisory group after Trump announced the withdrawal of the United States from the Paris Accords, but the reality is that the cost of Musk's ecological dream is far higher than he and others are willing to admit. Further reading: 'Cost of Elon Musk's Dream Much Higher than He and Others Imagine', *RealClearEnergy*, 8 June 2017.

10 Interview with John Petersen, 2016. See also 'How Large Lithium-ion Batteries Slash EV Benefits', 2016. A collection of Petersen's articles is available on the *Seeking Alpha* website.

11 See '*Les potentiels du vehicle électrique*' ['The Potentials of the Electric Vehicle'], *ADEME*, April 2016, and Troy R. Hawkins, Bhawna Singh, Guillaume Majeau-Bettez, and Anders Hammer Strømman, 'Comparative Environmental Life Cycle Assessment of Conventional and Electric Vehicles', *Journal of Industrial Ecology*, 17(1): 53–64, 2012.

12 Xinyu Chen, Hongcai Zhang, Zhiwei Xu, Chris P. Nielsen, Michael B. McElroy, and Jiajun Lv, 'Impacts of Fleet Types and Charging Modes for Electric Vehicles on Emissions under Different Penetrations of Wind Power', *Nature Energy*, 30 April 2018. For a more in-depth analysis of the environmental impact of an electric vehicle, refer to my article 'Do We Really Want Electric Vehicles?' in the English-language edition of the monthly paper *Le Monde Diplomatique*, September 2018.

13 To delve deeper into these questions, read the fascinating article by Jean-Marc Jancovici, 'Is the Electric Car an Ideal Solution for Tomorrow's Mobility?', 1 August 2015, on Jean-Marc Jancovici's website jancovici.com. See also 'Do We Really Want Electric Vehicles?', *Le Monde Diplomatique*, September 2018.

14 Rifkin, *The Third Industrial Revolution*, op. cit.

15 Rifkin, *The Zero Marginal Cost Society: the internet of things, the collaborative commons, and the eclipse of capitalism*, Palgrave Macmillan, 2014.

16 'US Car Sharing Service Kept 28 000 Private Cars Off the Road in 3 Years', *The Guardian*, 23 July 2016.

17 Eric Schmidt and Jared Cohen, *The New Digital Age: reshaping the future of people, nations and business*, Knopf, Random House Inc., 2013.

18 Imagine the staggering amounts of digital data this already generates. 'Every two days now we create as much information as we did from the dawn of civilization up until 2003,' say Schmidt and Cohen. This outlook has an economic impact: the 'sixth continent' that is the internet represents 22.5 per cent of the global economy and is expected to reach 25 per cent by 2020, or $24,000 trillion in revenues. Further reading: Mark Knickrehm, Bruno Berthon and Paul Daugherty, 'Digital Disruption: the growth multiplier — optimizing digital investments to realize higher productivity and growth', *Accenture Strategy*, 2016.

19 Every minute around the world, 2,400 trees are felled, which is

the equivalent of a third of the land area of France every year. See '*Déforestation: 18 millions d'hectares de forêts perdus en 2014*' ['Deforestation: 18 million hectares of forest lost in 2014'], *Le Monde*, 3 September 2015. 'In the tropical domain, net annual loss of forest area from 2000 to 2010 was about 7 million hectares, and net annual increase in agricultural land area was more than 6 million hectares'; 2016 State of the World's Forests, *Food and Agriculture Organisation of the United Nations*, 2016.

20 Fabrice Flipo, Michelle Dobré, and Marion Michot, *La Face cachée du numérique. L'impact environnemental des nouvelles technologies* [*The Dark Side of Digital: the environmental impact of new technologies*], L'Échappée, 2013.

21 Ibid.

22 Coline Tison and Laurent Lichtenstein, *Datacenter: hidden pollution*, Documentary, Camicas Productions, 2012. See also 'Email Miles — where does your email really go when you press send?', *The Huffington Post*, 18 February 2014.

23 The documentary takes us as far as the coalmines in Appalachia, in West Virginia, where the fossil fuel resources used to run the US power stations are extracted. 'There's nothing virtual about our clicks,' states the documentary. Referring to the illusion of dematerialisation, it poses the question: 'will our emails ultimately destroy the Appalachia mountains?'

24 Mark P. Mills, 'The Cloud Begins with Coal: big data, big networks, big infrastructure, and big power — an overview of the electricity used by the global digital ecosystem', August 2013.

25 'How Clean Is Your Cloud?', *Greenpeace*, April 2012.

26 'Elon Musk's SpaceX Is Striving to Win the Race to Build the Internet in Space', *Washington Post*, 15 May 2019.

27 As an example, in 1951, forty-four UNIVAC I computers (Universal Automatic Computer I) — the first US commercial computer — were sold. In 2015, nearly 300 million computers and over 200 million tablets were sold worldwide. Today, over half of the world's population possesses a mobile phone.

28 'It's less about scarce resources and more about "full dustbins",' write experts PierreNoël Giraud and Timothée Ollivier in *Économie des matières premières* [*The Economy of Commodities*], La Découverte, coll. Repères, 2015.

29 C.P. Baldé, V. Forti, V. Gray, R. Kuehr, and P. Stegmann, 'The Global E-waste Monitor – 2017, United Nations University (UNU), International Telecommunication Union (ITU) & International Solid Waste Association (ISWA)', Bonn/Geneva/Vienna.

30 The tally of electronic waste in 2017 was over 50 million tonnes versus 41 million tonnes in 2014. Further reading: 'Waste Crime – Waste Risks: gaps in meeting the global waste challenge', United Nations

Environment Programme (UNEP), 2015.

31 'Recycling Rates of Metals: a status report', United Nations Environment Programme (UNEP), 2011. These figures remain relevant, no subsequent report having been published.

32 Read '*La guerre des terres rares est déclarée*' ['The Rare-earth Metals War Is Declared'], *Terra Eco*, 19 April 2012. In addition, watch our documentary *Terres rares, le trésor caché du Japon*, Mano a Mano [*Rare Earths, Japan's Hidden Treasure*], 2012 (French only). Read also 'Global E-waste to Hit 49.8M Tons by 2018 — Here's what Japan is doing to combat it', *Forbes*, 23 November 2017.

33 Cerium, for instance — a rare earth used to polish glass — can be replaced with zirconium.

34 New uses have made it possible to reduce the quantity of europium and terbium in fluorescent lamps by 80 per cent, and the quantity of dysprosium used in magnets by 30 per cent. European car manufacturers are even looking for ways to make magnets without the need for rare-earth metals.

35 Interview with Jack Lifton, Technology Metals Research, 2016.

36 Ibid.

37 Surendra M. Gupta, *Reverse Supply Chains: issues and analysis*, CRC Press, 2013. Read also Rémy Le Moigne's fascinating '*L'Économie circulaire: comment la mettre en oeuvre dans l'entreprise grâce à la reverse supply chain?*' [*Using the Reverse Supply Chain to Implement the Circular Economy in Business*], Dunod, 2014.

38 This is the case for all commodity prices. The end of 2014 marked the end of a super cycle during which commodity prices were at their highest. The commodity markets have since bottomed out.

39 They are: aluminium, cobalt, chrome, copper, gold, iron, lead, manganese, niobium, nickel, palladium, platinum, rhenium, rhodium, silver, tin, titanium, and zinc.

40 Magnesium, molybdenum, and iridium.

41 Ruthenium, cadmium, and tungsten.

42 All these figures can be found in the 2011 United Nations Environment Programme (UNEP) publication 'Recycling Rates of Metals: a status report'. The rate of recycling of certain rare metals have decreased even further since 2011, the year in which this report was published. This is due to the drop in metals prices, which had made recycling less financially attractive. The rate of recycling of rhenium is virtually nonexistent since the last recycling company that processes the metal closed its operations in early 2018. Interview with Vincent Donnen, cofounder of Compagnie Des Métaux Rares, 2019.

43 Communication from the Commission to the European Parliament, the Council, the European Economic and Social Committee, and the Committee of the Regions on the 2017 list of Critical Raw Materials

for the EU.

44 'Hitachi Recycling Scarce Rare Earths', *The Japan Times*, 10 December 2010.

45 Interview with Christian Thomas, founder of Terra Nova Développement, 2017.

46 Basel Convention on the Control of Transboundary Movements of Hazardous Wastes and Their Disposal, adopted in Basel on 22 March 1989.

47 In particular, the convention bans waste containing hexavalent chromium compounds, copper, zinc, cadmium or antimony.

48 'EU Serious and Organised Crime Threat Assessment (SOCTA)', Europol, 2013.

49 J. Huisman, I. Botezatu, L. Herreras, M. Liddane, J. Hintsa, V. Luda di Cortemiglia, P. Leroy, E. Vermeersch, S. Mohanty, S. van den Brink, B. Ghenciu, J. Kehoe, C.P. Baldé, F. Magalini, and A. Bonzio, 'Countering WEEE Illegal Trade (CWIT) Summary Report, Market Assessment, Legal Analysis, Crime Analysis and Recommendations Roadmap', 31 August 2015, Lyon, France.

50 Martin Eugster and Roland Hischier, 'Key Environmental Impacts of the Chinese EEE-Industry', Tsinghua University, China, 2007.

51 See '*Les terres rares: des propriétés extraordinaires sur fond de guerre économique*' ['Rare-earth Metals: extraordinary properties against a backdrop of economic war'] with Paul Caro, rare-earth metals expert at the Académie des sciences. See also 'China Tries to Clean Up Toxic Legacy of its Rare Earth Riches', *New York Times*, 22 October 2013.

52 'Rare Earth and Radioactive Waste: a preliminary waste stream assessment of the Lynas advanced materials plant, Gebeng, Malaysia', National Toxics Network, April 2012. Likewise, in Malaysia, where our investigation led us in 2011: between the end of the 1970s and 1994, the Japanese multinational corporation Mitsubishi mined and refined rare-earth metals in Bukit Merah, in the north of the country. As explained to us by environmental activist Tan Ka Kheng, 'Their activities generated tremendously high levels of radiation. This in the nuclear industry is considered medium-activity radioactive waste and must therefore be manoeuvred with the greatest of care. But … they disposed of the waste in old, rusty barrels and used plastic bags as cladding. What a tragedy! Mitsubishi then closed the facility and left! Now we're left with this waste for 14.4 billion years!' Bukit Merah is one of the most radioactive sites in Asia. While we were there, $100 million were committed to renovate the site. Watch the author's and Serge Turquier's documentary *Rare Earths: the dirty war*, 2012.

53 'The Growing Role of Minerals and Metals for a Low Carbon Future', *World Bank Group*, June 2017.

54 'Electrification Puts the Car Industry at Risk' Says PSA Boss Tavares

at Frankfurt', *The Telegraph*, 12 September 2017. Carlos Tavares also stated that 'All this agitation and chaos will come back to haunt us because we will have made the wrong decisions that are insufficiently considered, lack perspective, and are made based on day-to-day emotions.' A year later, Yoshihiro Sawa, president of the luxury carmaker Lexus, warned against electric vehicles: 'EVs currently require a long charging time and batteries that have an environmental impact at manufacture and degrade as they get older. And then, when battery cells need replacing, we have to consider plans for future use and recycling. It's a much more complex issue than the current rhetoric perhaps suggests. I prefer to approach the future in a more honest way.' See 'Lexus Boss on EVs, Autonomy and Radical Design', *Autocar*, 11 August 2018.

Chapter Three: Delocalised pollution

1 For further reading, see Philippe Bihouix and Benoît de Guillebon's *Quel futur pour les métaux?*, op. cit.
2 See Mineral Resources Online Spatial Data, U.S. Geological Survey.
3 'Sustainable management of mineral raw materials': information report by the Sustainable Development Commission of the French National Assembly, 2011.
4 Ibid.
5 See 'Rare Earth Mining at Mountain Pass', *Desert Report*, March 2011.
6 Interview with Chen Zhanheng, deputy secretary general of the Association of China Rare Earth Industry, 2016.
7 Interview with Eric Noyrez, former CEO of Lynas and current managing director of Serra Verde, 2019.
8 'The global economic crisis put the project on hold for over a year. In September 2009, Nicholas Curtis managed to raise enough capital to revive the project, and engineering works began in January 2010. In the end, the Mount Weld mine produced its first gram of rare earths in 2013,' adds Noyrez.
9 Interview with Jean-Yves Dumousseau, the then sales director of US chemicals company Cytec, 2011.
10 Nicolas Hulot later became the environment minister under the Macron presidency.
11 The full history of the group is available on the Solvay website (French only).
12 Interviews with Jean-Paul Tognet, former industrial and raw materials director at Rhône-Poulenc and Rhodia Terres Rares, 2016 and 2017.
13 The shore of Port Neuf lies opposite the Minimes marina, less than 2 kilometres from the Saint Nicolas tower.
14 Interview with Bruno Chareyron, nuclear physics engineer at CRIIRAD, 2016.

15 Alain Roger and François Guéry (dir.), *Maîtres et protecteurs de la nature* [*Masters and Protectors of Nature*], Champ Vallon, 1991. According to Régis Poisson, a former engineer at Rhône-Poulenc, one day the local authorities apparently asked: 'Can't you take the red out of the plumes to make them look less dirty?'
16 '*La CRIIRAD crie à la radioactivité dans la baie de La Rochelle*' ['CRIIRAD Exposes Radioactivity in the Bay of La Rochelle'], *Libération*, 19–20 March 1988.
17 The problem appears to have lasted until at least 2002, when a new report from the CRIIRAD demonstrated 'the ongoing contamination of the old outfall pipe' on the shore of Port Neuf. Refer to CRIIRAD report no. 10–149 V1 1, '*Mesures radiamétriques sur terrain de l'université de La Rochelle*' ['Radiometric In-field Measurements by La Rochelle University'], 15 December 2010. According to Bruno Chareyron, 'The facility had not been dismantled and the immediate environment of the outfall pipe wasn't even cordoned off. The company had not done everything in its power to limit local residents' exposure to radiation.'
18 '*La CRIIRAD crie à la radioactivité dans la baie de La Rochelle*' ['CRIIRAD Exposes Radioactivity in the Bay of La Rochelle'], op. cit.
19 Interviews with Jean-Paul Tognet, 2016 and 2017.
20 Jean-Yves Le Déaut, 'Report on Low-level Radioactive Wastes', *Parliamentary Office for Science and Technology Assessment (OPECST)*, 1992.
21 Interviews with Jean-Paul Tognet, 2016 and 2017.
22 Ibid. I contacted Jean-René Fourtou, but he did not respond to my request for an interview.
23 Interview with Jean-Yves Dumousseau, 2016. Jean-Paul Tognet states that the Chinese prices are around 25 per cent lower than those of the competition.
24 Ibid.
25 'Toxic Memo', *Harvard Magazine*, 5 January 2001.
26 Interview with Patrice Christmann of BRGM, the French Geological Survey, 2013.
27 Regulation (EC) No 1907/2006 of the European Parliament and of the Council of 18 December 2006 concerning the Registration, Evaluation, Authorisation and Restriction of Chemicals (REACH), establishing a European Chemicals Agency.
28 Harvey Black, 'Chemical Reaction: the U.S. response to REACH', *Environmental Health Perspectives*, March 2008.
29 Interview with Christophe-Alexandre Paillard, Deputy Director of the Strategic Affairs Directorate (DAS, French Ministry of Defence), 2013.
30 As rightly put by Louis Maréchal, currently extractives sector policy adviser within the OECD's Responsible Business Conduct Unit, by

transferring the responsibility of production to mining countries we have also relocated the associated societal repercussions: corruption, conflict, governance issues, the black market, human rights violations, and so on. Interview with Louis Maréchal, 2017.

31 Keynote by Gregory Bowes, Chairman of Northern Graphite, 5th Annual Cleantech & Technology Metals Summit: Invest in The Cleantech Revolution, 2016.

32 Interview with Xue Lan, professor of political science at Tsinghua University, 2016.

33 'Annual Report Pursuant to Section 13 or 15(d) of the Securities Exchange Act of 1934 for the Fiscal Year Ended September 29, 2018.', Apple Inc, 2018. Apple's 2019 environmental report mentions the presence of metals such as rare earths, cobalt and tungsten in their mobile phones, but omits the conditions under which they are extracted and refined. 'Environmental Responsibility Report, 2019 Progress Report, Covering Fiscal Year 2018', Apple Inc, 2019.

34 'Environmental Impact Report', Tesla Motors, Inc., 2019.

35 To give just one example, read 'Time to Recharge: corporate action and inaction to tackle abuses in the cobalt supply chain', Amnesty International, 2017.

36 'The Fairphone, a No-conflict Smartphone without Planned Obsolescence', *World Forum for a Responsible Economy*, 28 July 2017.

37 'The "Right to Repair" Movement Wants You to Be Able to Fix Your Own Stuff', *Public Radion International*, 24 December 2018.

38 Visit The Restart Project website.

39 'Meet the $21 Million Company that Thinks a New iPhone Is a Total Waste of Money', *Inc*, April 2017.

40 Under this framework, by 2030 member states are required to reduce their CO_2 emissions by 40 per cent from 1990 levels, and increase the share of renewables energies to 27 per cent of their consumption.

41 'U.S. Leads in Greenhouse Gas Reductions, but Some States Are Falling Behind', *Environmental and Energy Study Institute*, 27 March 2018.

42 This reality certainly helps to explain a discussion that 'fell on deaf ears' between Song Xi Chen, economics professor at Peking University, and a US steel trader from Milwaukee, both travelling on a plane from Chicago to Beijing in 2014. 'I said to him: "You are responsible for the pollution in China!" To which he replied: "But I don't own the industry!"' Interview with Song Xi Chen, 2016.

43 US military spending steadily declined over the 1990s until 11 September 2001. See 'Trends in U.S. Military Spending', Council of Foreign Relations, 15 July 2014. This is a trend reflected in France, where the defence budget, calculated in constant euro terms, fell 20 per cent in the space of twenty-five years to €31.4 billion in 2015. See

'En euros constants, le ministère de la Défense a perdu 20 per cent de son budget en 25 ans' ['In Constant Euro Terms, the Ministry of Defense Budget Has Fallen 20 per cent in 25 Years'], *Le Monde*, 29 April 2015.

44 Interview with Alain Liger, former secretary-general of the Comité pour les métaux stratégiques (COMES – Committee for Strategic Metals), 2016.

45 Interview with Chris Ecclestone, founder of Hallgarten & Company, 2016. According to Ecclestone, the stockpiles were not managed by the US Treasury, but by the Pentagon. By selling them off, the US Army was able to offset the fall in military lending and buy new equipment, such as drones, aircraft, and guided bombs.

46 Interview with Jean-Philippe Roos, then commodity market analyst in the Natexis Asset Management economic research department, 2010. These transactions apparently began even earlier: after signing the treaties resulting from the Strategic Arms Limitation Talks (SALT) I and II, the USSR dismantled some of its atomic bombs and sold the uranium to the US. The massive quantity of minerals that suddenly flooded the market also contributed to 'killing' the US uranium industry.

47 *'Braderie forestière au pays de Colbert'* ['The Sell-off of the Forestry Industry in the Country of Colbert'], *Le Monde diplomatique*, October 2016.

48 *'Grasse se remet au parfum'* ['Grasse Back on the Scent'], *M. Le magazine du Monde*, 11 July 2016.

49 Thales registration document, 2015.

50 Interview with Alain Liger, 2016.

51 In Portugal, the BRGM geologists discovered the Neves-Corvo copper orebody. In Quebec, their exploration activities enabled the mining of copper and zinc orebodies, the Langlois mine.

52 Interview with Alain Liger, 2016.

53 Ibid.

54 'Trends in the Mining and Metals Industry', ICMM (International Council of Mining & Metals), October 2012.

55 Jean-Marie Guéhenno, French White Paper on Defence and National Security 2013, Ministry of Defence, 2013.

56 Upon his retirement in 1984, Alain de Marolles wrote a predictive analysis in which he referred to a 'third industrial revolution' in progress, spurred by the electronics industry, the space race biology. He commented on the massive demand for metals generated by these sectors, and predicted that some of these metals — copper, cobalt, manganese, nickel, platinum, and gold — would in the future be mined from the bottom of the sea. See Alain de Gaigneron de Marolles, *L'Ultimatum: fin d'un monde ou fin du monde?* [*Ultimatum: end of one world or end of the world?*], Plon, 1984.

57 Interview with Jack Lifton, 2016.

58 Interview with Didier Julienne, natural resources strategist, 2016.

59 '50 Years Ago: cargo cults of Melanesia', *Scientific American*, 1 May 2009.
60 'The Surprising Number of American Adults Who Think Chocolate Milk Comes from Brown Cows', *The Washington Post*, 15 June 2017.

Chapter Four: The West under embargo

1 Mineral Commodity Summaries, *U.S. Geological Survey*, 2017.
2 Communication from the Commission to the European Parliament, the Council, the European Economic and Social Committee, and the Committee of the Regions on the 2017 list of Critical Raw Materials for the EU.
3 Interview with Felix Preston, energy, environment and resources specialist at Chatham House, 2016. See also Felix Preston, Rob Bailey and Siân Bradley (Chatham House), 2016, and Dr Wei Jigang and Dr Zhao Changwen (DRC), 'Navigating the New Normal: China and global resource governance', January 2016.
4 See the information report on the sustainable management of mineral commodities for the French Committee on Sustainable Development and Planning, presented by MPs Christophe Bouillon and Michel Havard, French National Assembly, 2011.
5 '*La Chine met les matières premières sous pression*' ['China Puts Raw Materials Under Pressure'], *Les Echos*, 7 July 2015. See also 'Metals Shine on China Demand', *China Daily*, 19 June 2018.
6 Interview with Andrew Peaple, journalist writing on commodities for the Hong Kong office of *The Wall Street Journal*, 2016.
7 Geneviève Barman and Nicole Dulioust, '*Les années françaises de Deng Xiaoping*' ['Deng Xiaoping's French Years'], *Vingtième Siècle. Revue d'histoire*, 1988, vol. 20, no. 1, pp. 17–34.
8 '*Le singe et la souveraineté des ressources*' ['The Monkey and the Sovereignty of Resources'], *Le Cercle – Les Échos*, 12 February 2016.
9 Information report no. 349 (2010-2011) on the security of France's strategic supplies, prepared for the French Committee on Foreign Affairs, Defence and Armed Forces by Jacques Blanc, 2011.
10 Generally speaking, Western countries took a position of reliance on major mineral-producing countries. Accordingly, the United States is 100 per cent dependent for seventeen minerals including rubidium, scandium, graphite, indium and thorium. It is 80 per cent dependent for twenty-nine metals and 50 per cent dependent for forty-one metals. See 'Going Critical: being strategic with our mineral resources', *USGS*, 13 December 2013. As for the EU, after a review of fifty-four metals, the findings from Brussels are more or less the same: 'around 90 per cent of global supply [of EU members] originated from extra-EU sources'. See 'Report on Critical Raw Materials for the EU', *Report of the Ad Hoc Working Group on Defining Critical Raw Materials*, May 2014.

11 Interview with Dudley Kingsnorth, professor at Curtin University in Australia, 2016.
12 Refer to the figures reported by John Seaman in 'Rare Earth and Clean Energy: analyzing China's upper hand', *Institut français des relations internationales* (IFRI), September 2010.
13 J. Korinek and J. Kim, 'Export Restrictions on Strategic Raw Materials and their Impact on Trade', OECD Trade Policy Papers, no. 95, OECD Publishing, 2010.
14 Refer to complaints DS295, DS395, and DS398 brought to the WTO by the United States, the European Communities, and Mexico with respect to China's application of export restrictions on various raw materials of which it is the origin. The Chinese rare metals strategy stems from the country's desire to re-establish its status as a global powerhouse. Propelled by its economy, which has grown over tenfold since the turn of the millennium, China is increasingly flexing its muscles by weighing in more on international affairs. Of note, this diplomatic activism led to the creation of the Asian Infrastructure Investment Bank (AIIB) in 2014 to counteract the hegemony of the International Monetary Fund (IMF). Similarly, Beijing has strengthened its bilateral relations with its regional neighbours. Since the start of the 2000s, China has acquired and built port facilities that extend from its coasts to Port Sudan in east Africa — the 'string of pearls' strategy aimed at containing its Indian neighbour. Not to mention the construction of artificial islands in the Spratly Archipelago, a disputed maritime region of the South China Sea known for its gigantic gas reserves. China has no qualms either about taking on Japan, robbing it of its position as the second-biggest global economic power in 2010, making China the economic leader in the region.
15 Interview with Jean-Yves Dumousseau, 2016.
16 See also Robert L. Paarlberg, 'Lessons of the Grain Embargo', *Foreign Affairs*, Fall 1980.
17 'Russia's Gas Fight with Ukraine', BBC, 31 October 2014.
18 See https://www.youtube.com/watch?v=nQRjZAvr8HI.
19 Interview with Toru Okabe, professor at the University of Tokyo, 2011.
20 'Amid Tension, China Blocks Vital Exports to Japan', *The New York Times*, 22 September 2010.
21 'The Difference Engine: more precious than gold', *The Economist*, 17 September 2010.
22 'Continental AG, Bosch Push EU to Secure Access to Rare Earths', Bloomberg, 1 November 2010.
23 Hilary Clinton added that 'our countries, and others will have to look for additional sources of supply. This served as a wake-up call.' 'Clinton Hopes Rare Earth Trade to Continue Unabated', *Reuters*, 28 October 2010.
24 Interview with John Seaman, researcher at the French Institute of

International Relations (IFRI), 2015.

25 A kilogram of terbium was soon trading at €2,900 — that is, ten times higher than two years before. Source: '*Rhodia renouvelle ses filons de terres rares*' ['Rhodia Renews Its Rare Earths Arm'], *L'Expansion*, 2 November 2011. In mid-2011, a kilogram of dysprosium traded at an astronomic price of nearly $3,000 — that is, 100 times more than in 2003. Source: '*Les matières premières comme enjeu stratégique majeur: le cas des "terres rares"*' ['Commodities as a Major Strategic Challenge: the case of rare earths'], presentation by Christian Hocquard. See http://archives.strategie.gouv.fr/cas/content/23e-rendez-vousde-la-mondialisation-matieres-premieres-metaux-rares-ressources-energetiques.html.

26 'Had a $2 cup of coffee undergone the same inflation as europium, today it would be worth $24.55,' explains the company General Electric to its customers to explain the increases in its tariffs. Source: '*Rhodia renouvelle ses filons de terres rares*', op. cit.

27 In 2016, Royal Bafokeng Holdings decided to reduce their share to 6.3 per cent in order to diversify their source of revenue. See 'Why Royal Bafokeng Is Selling Implats', *Moneyweb*, 5 April 2016.

28 Thanks to their revenue, the Bafokeng have embarked on a large-scale investment campaign in a variety of sectors, including insurance, telecommunication, sports, and infrastructure. To 'crown' it all, they have designed a plan called Vision 2035 — the year in which platinum mining will have peaked. Thereafter, the foundations of an independent and sustainable model for the platinum economy will need to be laid.

29 At a regional level, chiefs from other communities in South Africa and Zambia, and even Democratic Republic of Congo (DRC) president Joseph Kabila, came.

30 From 2012, the Mongolian government promulgated a law that drastically restricts foreign investment in sectors — such as mining — that are seen as strategic.

31 'Qatar Fund Raises Stake in Xstrata', *The New York Times*, 23 August 2012. Xstrata merged with Glencore in 2013.

32 'Rising Resource Nationalism Seen as "Fire Burning" for Miners', *Bloomberg*, 16 May 2018, and 'Resource Nationalism on the Rise in Sub-Saharan Africa', *Mining Weekly*, 15 June 2018.

33 Interview with Sukhyar Raden, General Director — Mineral & Coal for the Indonesian Ministry of Energy & Mineral Resources, 2015.

34 'Export Restrictions in Raw Materials Trade: facts, fallacies and better practices', OECD, 2014.

35 These claims were in fact supported by Resolution 1803 (XVII) of the United Nations, voted in 1962, which established 'permanent sovereignty of peoples and nations over their natural wealth and resources'.

36 Interview with Jack Lifton, 2016.

37 The protectionism is also essentially the outcome of Western trade practices. China, too, is the target of trade retaliation measures orchestrated by Western states and corporations. Since 2011, protectionist measures against China have tripled, according to the Global Trade Alert (GTA) in its report 'The Global Trade Disorder' (2014). The financial crisis triggered by the collapse of Lehman Brothers was unquestionably a turning point. States went from their traditional role of safeguarding international trade to becoming one of its harshest critics. In this context, 'China has to manage a shift in the globalisation paradigm at a time when the former winners thereof — the West — are in the process of becoming the losers', says Brian Jackson, an analyst at the Beijing bureau of IHS group. This has the World Economic Forum worried. Recognising the multiplication of attacks on the trade in mining resources, the Swiss forum is deeply concerned about scenarios which, in one of its publications, it codes 'amber' or even 'red'. This latest prediction depicts a world in which 'markets are shaped by state interventionism', and 'trade is defined by a complex web of protectionist barriers and preferential agreements', and by 'limited cross-border flows of products, labour and capital'. See the report *Mining & Metals: scenarios to 2030*, World Economic Forum, 2010.

38 Global steel production stood at 1,808 million tonnes in 2014. See 'Global Crude Steel Output Increases by 4.6 Per Cent in 2018', World Steel Association, 25 January 2019.

39 To date, China has not signed the Global Reporting Initiative, the purpose of which is to encourage the transparency of governments, particularly with regard to how they manage their resources.

40 *'Matières premières: le grand retour des stratégies publiques'* ['Raw Materials: state strategies make a big comeback'], *Paris Technology Review*, 4 May 2012.

41 For further understanding of the growing role of the financial sector on the commodities market, see the documentary *Commodity Traders* by Jean-Pierre Boris and Jean Crépu (2014). The following is an extract transcribed from the documentary: 'March 13th, 2000. The dotcom bubble bursts. The Nasdaq stock market index crashes. Financiers pull out of this market in search of new sources of profit. Two US economists, Gary Gorton and K. Geert Rouwenhorst, take a closer look and publish a report entitled "Facts and Fantasies about Commodity Futures". It underscores the high profitability of investing in commodities, which is also a way to diversify share portfolios. The major banks got the message and turned their sights on the commodities market.'

42 There have also been cases of speculation on palladium, cobalt, and molybdenum.

43 'Electric Carmakers on Battery Alert after Funds Stockpile Cobalt',

The Financial Times, 23 February 2017. Another example is the massive stockpiling of indium at the end of 2009 via commodity investment vehicles to 'dry up this small market and massively inflate prices', say the press. See '*La Chine restreint ses exportations de matières premières stratégiques*' ['China Limits Its Strategic Raw-material Exports'], *Le Monde*, 29 December 2009.

Chapter Five: High-tech hold-up

1 The metals must be smelted in precise amounts. The resulting alloy is then cooled, milled, compacted, pressed using a piston, sintered, and cooled again. For a more technical explanation, read Sandro Buss, '*Des aimants permanents en terres rares*' ['Permanent Rare-earth Magnets'], *La Revue polytechnique*, no. 1745, 13 April 2010.

2 Turning the sparkwheel of the lighter strikes the flint — a blend of rare earth metals called 'mischmetal' — producing a spark that lights the gas or the wick. As for camping lanterns, the incandescent light of the flame is produced not by gas but by the rare earth cerium. When heated by the flame, the cerium-impregnated gas mantle emits a bright white light that provides enhanced lighting. Refer to '*Terres rares: enjeux stratégiques pour le développement durable*' ['Rare Earths: strategic challenges for sustainable development'], a talk by Patrice Christmann, deputy director of corporate strategy at the French Geological Survey (BRGM), as part of a series of seminars organised by the National Scientific Research Council (CNRS) of France, 17 September 2013.

3 The rare earths covering the inner face of the screen are 'excited' by the cathode-ray tube to emit coloured light, and therefore images. Europium reproduces the colour red, and terbium the colour green.

4 These are the samarium-cobalt (Sm-Co) magnets, the chemical formulae of which are $SmCo_5$ and Sm_2Co_{17}, and magnets made using the rare earth neodymium, iron and boron (NdFeB), the chemical formula of which is $Nd_2Fe_{14}B$. They were invented by Masato Sagawa of Sumitomo Special Metals in Japan, and John Croat from General Motors in the US.

5 For example, samarium 'is used mainly to make permanent magnets. A technology used for the new generation of Alstom's TGVs [high-speed trains], and enabling the production of engines 30 to 40 per cent more compact, and 10 to 20 per cent more powerful.' See '*Le CAC 40 accro aux "terres rares"*' ['The CAC 40's Addiction to "Rare Earths"'], *L'Expansion*, 12 November 2012.

6 Interview with Jack Lifton, 2016. According to Philippe Degobert, electrical engineering lecturer at the École nationale supérieure des Arts et des Métiers (ENSAM), and director of Masters in Mobility and Electric Vehicles (MVE), the ferrite magnet is seven times weaker than a samarium magnet, and ten times weaker than a neodymium magnet.

7 Interview with Chen Zhanheng, 2016.
8 Interview with Chris Ecclestone, 2016.
9 See Régis Poisson, '*La guerre des terres rares*' ['The Rare Earths War'], *L'Actualité chimique*, no. 369, December 2012.
10 Interviews with Jean-Paul Tognet, 2016 and 2017.
11 Ibid.
12 Interview with Jean-Yves Dumousseau, 2016.
13 Régis Poisson, '*La guerre des terres rares*', op. cit.
14 Interview with Jean-Yves Dumousseau, 2016.
15 Interviews with Jean-Paul Tognet, 2016 and 2017.
16 Ibid.
17 Ibid.
18 Interview with Jim Robinson, United Steelworkers (USW), 2011.
19 Interview with Régis Poisson, former engineer at Rhône-Poulenc, 2013.
20 See also Régis Poisson, '*La guerre des terres rares*', op. cit.
21 Interview with David Merriman, Roskill Consultancy, 2016.
22 According to a French industrialist who preferred to speak anonymously, the purchase price ratio for European producers and Chinese producers can be as high as 1:7 — a price differential that Jean-Paul Tognet finds excessive.
23 Interview with Jean-Yves Dumousseau, 2011.
24 See 'Baotou Rare Earth High-Tech Industrial Development Zone', *China Daily*, 27 October 2015.
25 Since our trip to Baotou, 'Inner Mongolia will present a broad platform for investment cooperation in fields such as new energies, advanced materials, energy saving, environmental protection, high-end equipment, big data cloud computation, bio technology, traditional Inner Mongolian medicine and "Tourism +".' See 'Central Enterprises Launch a New Wave of Investments in Inner Mongolia, Signing Contracts Worth RMB 402 Million Yuan at an Informal Meeting', *PR Newswire*, 1 March 2017. The city of Baotou is still pursuing its ambitions by developing an industrial area of 470 square kilometres dedicated to rare earths. See 'Huge Rare Earth Industrial Park Coming to Inner Mongolia', *China Daily*, 29 August 2017.
26 Interview with Dudley Kingsnorth, 2016.
27 Industrial robots are in fact the cornerstone of tomorrow's smart and ultra-connected '4.0 factories', where Germany has the lead. In 2015, the German machine tool industry generated over €15 billion, exported three-quarters of its production and had a headcount of nearly 70,000 employees. It is a pillar of the national economy. See 'German Machine Tool Industry Expects Moderate Growth in 2016', *Verein deutscher Werkzeugmaschinenfabriken*, 2016.
28 Interview with Chris Ecclestone, 2016.
29 For a more detailed explanation, see the French Geological Survey's

public report: '*Panorama du marché du tungstène*' ['View of the Tungsten Market'], BRGM, July 2012.

30 Interview with Chris Ecclestone, 2016.

31 The Mittelstand may have won the battle, but they did not win the war. China has a keen interest in some of Germany's star industrial robots, such as KUKA. See '*Allemagne: le "Mittelstand" face à l'offensive chinoise*' ['Germany: the Mittelstand faces the Chinese offensive'], *Le Monde*, 4 June 2016.

32 Graphene's applications are astounding: bendy mobile phones, see-through laptops, ultra-powerful nanoprocessors, or nanochips that can be inserted into the human body to detect cancers, and so on.

33 'U.S. Brings WTO Challenge against China over Copper, Graphite, Other Minerals', *The Wall Street Journal*, 13 July 2016.

34 'United States Expands Its Challenge to China's Export Restraints on Key Raw Materials', Office of the United States Trade Representative, July 2016.

35 Interview with Daisy Chen, journalist at the Beijing bureau of Metal Pages, 2016. The case is ongoing and can be consulted on the WTO website (DS558: China — Additional Duties on Certain Products from the United States).

36 Interview with Chris Ecclestone, 2016.

37 Interview with Thomas Kruemmer, 2016.

38 And why not do an inventory of 'critical super-magnets' that could potentially be in short supply? It's a legitimate question since China, as part of its strategy to attract industries that use rare earths, threatened Denmark, a major producer of wind turbines, to suspend its rare-earth magnet exports, according to Chris Ecclestone.

39 Interview with Didier Julienne, 2016.

40 Interview with Jack Lifton, 2016.

41 Interview with Alain Liger, 2016.

42 'Tin: the secret to improving lithium-ion battery life', *Forbes*, 23 May 2012.

43 Interview with Agung Nugroho Soeratno, head of communication at PT Timah, 2014.

44 Interview with Peter Kettle, analyst at the International Tin Research Institute, in the UK, 2016.

45 'Shanghai to Match London Metals as China Seeks Commodities Sway', *Bloomberg News*, 26 March 2015. Futures are one of the financial instruments traded on the derivatives market. To offset price instability, buyers and sellers agree on the future sale of goods based on a price set in advance.

46 'China ShFE Plans Commodities Platform to Set Physical Prices', *Reuters*, 15 May 2018.

47 'Bursa Malaysia Derivatives Introduces Futures Tin Contract', *The Star*, 29 September 2016.

48 There is no guarantee that such a policy had any bearing on global prices in the short term. Peter Kettle nevertheless maintains that 'there are clear opportunities for local brokers who benefit from financial transactions executed locally that would otherwise have taken place in the UK'. As such, stock exchanges like these actually strengthen the status of Asian cities as financial hubs.
49 Nickel and bauxite. See 'Indonesia Eases Ban on Mineral Exports', *The Financial Times*, 13 January 2017.
50 Hervé Kempf, *Fin de l'Occident, naissance du monde* [*End of the West, Birth of the World*], Seuil, 2013.
51 From a speech given at a conference organised by Cyclope Circle in Paris 2016.
52 See the African Union's February 2009 'Africa Mining Vision' report.
53 Refer to the World Bank Report 'Increasing Local Procurement by the Mining Industry in West Africa', January 2012.

Chapter Six: The day China overtook the West

1 Claude Chancel and Libin Liu Le Grix, *Le Grand Livre de la Chine* [*The Big Book of China*], Eyrolles, 2013.
2 James McGregor, APCO senior counsellor, describes China's National Medium- and Long-Term Plan for the Development of Science and Technology (MLP) as the 'grand blueprint of science and technology development' to bring about the 'great renaissance of the Chinese nation'. James McGregor, China's Drive for 'Indigenous Innovation' —A Web of Industrial Policies, U.S. Chamber of Commerce, 27 July 2010.
3 These industries are energy conservation and environmental protection, next-generation IT, biotechnology, high-end equipment manufacturing, new energy, new materials, and clean-energy vehicles. See 'China's 12th Five-Year Plan: how it actually works and what's in store for the next five years', APCO Worldwide, 10 December 2010.
4 Refer to the summary of the 13th Five-Year Plan in 'Prosperity for the masses by 2020 – China's 13th Five-Year Plan and its business implications', PwC China, Hong Kong and Macau, 2015.
5 Interview with Xue Lan, 2016.
6 I have borrowed this breakdown from Malo Carton and Samy Jazaerli's book, *Et la Chine s'est éveillée. La montée en gamme de l'industrie chinoise* [*And Then China Stirred: Chinese industry moves upmarket*], Presses de l'École des mines, 2015.
7 Interview with Ding Yifan, researcher at the Institute of World Development, 2016.
8 The National Medium- and Long-Term Plan for the Development of Science and Technology (2006–2020), The State Council of the People's Republic of China, 2006.

9 James McGregor, 'China's Drive for "Indigenous Innovation" — A Web of Industrial Policies', op. cit.
10 Ibid.
11 See Jean-Louis Beffa, *Les Clés de la puissance* [*The Keys to Power*], Seuil, 2015.
12 The full name is '1986 National High Technology Research and Development Program'. The first two numbers of Program 863 refer to the year the program was started, and the third number to the month (March).
13 These industries are information technology, biology, aerospace, automation, energy, materials, and oceanography.
14 '2019 Global R&D Funding Forecast', *R&D Magazine*, Winter 2019.
15 Interview with Bo Chen, professor at the Shanghai Free Trade Zone research institute, 2016.
16 'China "Employs 2 Million to Police Internet"', *CNN*, 7 October 2013.
17 For a critique of the innovation ecosystem in the Chinese mineral resources sector, see the report by Nicholas Arndt (Institute of Earth Sciences), Thierry Augé (French Geological Survey) and Michel Cuney (Geo-Resources Laboratory of Université de Lorraine), '*Les Ressources minerals en Chine*' ['Mineral Resources in China'], July 2014.
18 Interview with Bruno Gensburger, 2016. One French expert I spoke to who preferred not to be quoted said he heard the following remark from the lips of the Chinese director of an electronics group in reference to his staff: 'They don't have ideas because they are obedient. And if they're not obedient, I hit them.'
19 Bernard Apremont, '*L'économie de l'URSS dans ses rapports avec la Chine et les démocraties Populaires*' ['The Economy of the USSR in Its Relations with China and Popular Democracies'], *Politique étrangère*, 1956, vol. 21, no. 5, pp. 601–13.
20 James McGregor, 'China's Drive for "Indigenous Innovation" — A Web of Industrial Policies', op. cit.
21 Interview with Julien Girault, journalist at the Agence France-Press Beijing desk, 2016.
22 Interview with Ding Yifan, 2016.
23 'China Will Attempt 30-plus Launches in 2019, Including Crucial Long March 5 Missions', *SpaceNews*, 29 January 2019.
24 Interview with Xue Lan, 2016. This progress heralded the growth in developing countries' contribution to the production of skills vis-à-vis the duopoly held by the US and Europe. A new 'duel of intelligence' (to use the expression coined by Claude Chancel and Libin Liu Le Grix in *Le Grand Livre de la Chine*, op. cit.) that Irina Bokova, the then director-general of UNESCO, highlighted in the 2015 UNESCO report on science. In it, she stated that 'the North–South divide in research and innovation is narrowing, as a large number of countries are incorporating science, technology and innovation in their national

development agendas.' See 'UNESCO Science Report: Towards 2030', 2015. Five years earlier, Bokova also stated: 'The bipolar world in which science and technology (S&T) were dominated by the Triad made up of the European Union, Japan and the USA is gradually giving way to a multi-polar world, with an increasing number of public and private research hubs spreading across North and South.' See 'Research and Development: USA, Europe and Japan increasingly challenged by emerging countries, says a UNESCO report', *Unescopress*, 10 November 2010.

25 'China Continues to Dominate Worldwide Patent Applications', *Engineering & Technology*, 4 December 2018. And in 2018, the 'Science & Engineering Indicators' report of the US National Science Foundation also found that with 426,000 publications, China has become the world's biggest producer of scientific papers. See 'Science & Engineering Indicators 2018 Report', National Science Board, 2018. See also 'China Declared World's Largest Producer of Scientific Articles', *Scientific American*, 23 January 2018.

26 See Jean-Louis Beffa, *Les Clés de la puissance*, op. cit.

27 See the June 2019 'Top 500 List' online. In 2018, Beijing was again overtaken by two new US super computers.

28 '*La Chine devient la première puissance informatique au monde*' ['China Becomes the World's Leader in IT'], *Le Figaro*, 21 June 2016.

29 'Clean Energy Investment Exceeded $300 Billion Once Again in 2018', *Bloomberg*, 16 January 2019. China plans to increase this investment to $360 billion in 2020: 'China Aims to Spend at Least $360 Billion on Renewable Energy by 2020', *The New York Times*, 5 January 2017.

30 '*Énergies renouvelables: 2015, année record pour les Investissements*' ['Renewable Energies: 2015, a record year for investment'], *Les Échos*, 18 April 2016.

31 'Obama Says China Rare-earths Case Is Warning for WTO Violators', *Bloomberg*, 13 March 2012.

32 '*Voiture électrique: quand la Chine nous électrocutera*' ['Electric Cars: the day China electrocutes us'], *Caradisiac.com*, 16 October 2017. As stated by PSA CEO Carlos Tavares at the Frankfurt International Motor Show in September 2017: 'For a century, the Chinese chased after the internal combustion engine by paying royalties to the West. Now that they have found the point of disruption, they are taking the lead in electric cars which for the next century is the symmetry of their experience of the last.'

33 See 'Red China's Green Crisis', *Le Monde diplomatique*, July 2017.

34 These are the key takeaways of the research by Karl Gerald Van den Boogaart from the Freiberg University of Mining and Technology. His work was presented at the Denver SME Critical Minerals Conference in 2014 and quoted by Dudley Kingsnorth at the 5th Annual Cleantech and Technology Metals Summit in Toronto in April 2016.

35 Dudley Kingsnorth, professor at Curtin University, Western Australia.
36 A highly approximate estimation values the production market at $1,800 billion in 2015.
37 Interview with Peter Dent of the Electron Energy Corporation, 2011.
38 Interview with Dudley Kingsnorth, 2016.
39 Ibid.
40 'Donald Trump Hails New Era of US Energy "dominance"', *The Financial Times*, 30 June 2017.
41 These political choices are nevertheless based on the false assumption according to which the US alone can direct new energy balances in one direction rather than another, whereas the Chinese now hold the cards. In addition, by refusing to tackle Beijing head-on, Washington has already admitted defeat.
42 Arnaud Montebourg, '*L'Europe ne peut plus être à ce point désinvolte sur la mondialisation*' ['Europe Cannot Continue to Be so Lackadaisical about Globalisation'], *Le Monde*, 26 October 2016.
43 Interview with Didier Julienne, 2015.
44 Interview with Gary Hubard, United Steelworkers, 2011.
45 The share decreased from 16.4 per cent to 12.4 per cent. Refer to the 'Industry in France' infographic on www.gouvernement.fr, 2 April 2015. Note, however, the strong figures from the French industrial sector in 2017. See '*La France recrée enfin des usines*' ['France Is Finally Recreating Factories'], *Le Monde*, 29 September 2017.
46 'Industry (Including Construction), Value Added (Per Cent of GDP)', The World Bank, 2019.
47 Jean-François Dufour, *Made by China: les secrets d'une conquête industrielle* [*Made in China: the secrets of an industrial conquest*], Dunod, 2012.
48 Brahma Chellaney, 'The Challenge from Authoritarian Capitalism to Liberal Democracy', *China-US Focus*, 6 October 2016.
49 The expression was suggested by Joshua Cooper Ramo in a 2004 academic paper of the Foreign Policy Centre entitled 'The Beijing Consensus'.
50 Zhao Tingyang, *The Tianxia System: an introduction to the philosophy of a world institution*, Jiangsu Jiaoyu Chubanshe, 2005.
51 Interview with Chen Zhanheng, 2016.

Chapter Seven: The race for precision-guided missiles

1 In nuclear power plants, samarium-149 is used to absorb neutrons and thus reduce the reactivity (the rate of fission) of nuclear fuel. Contrary to the fiction, it is not a rare earth in itself, but an isotope of samarium.
2 Numerous reports look into the importance of rare earths in the US defence industries, such as Valerie Bailey Grasso's report 'Rare Earth Elements in National Defense: background, oversight issues, and options for Congress', *Congressional Research Service*, 23 December 2013.

3 Jean-Claude Guillebaud, *Le Commencement d'un monde. Vers une modernité Métisse* [*The Beginning of a World: towards hybrid modernity*], Seuil, 2008.
4 Interview with Jack Lifton, 2016.
5 Interview with David Merriman, 2016. David Merriman had less to say on the nature of these rare earths. Jean-Paul Tognet believes that the rare earth in question is scandium.
6 'One Budget Line Congress Can Agree On: spending billions on the US military', *The Conversation*, 14 August 2019.
7 For the full story on the delocalisation of Magnequench, read Charles W. Freeman III, 'Remember the Magnequench: an object lesson in globalization', *The Washington Quarterly*, The Center for Strategic and International Studies, 2009, pp. 61–76.
8 'Chinese Defence Spending to Grow 7.5 Per Cent in 2019 as Beijing Seeks "World-class" Military', *Japan Times*, 5 March 2019.
9 See Thierry Sanjuan, '*L'Armée populaire de libération: miroir des trajectoires modernes de la Chine*' ['The People's Liberation Army: mirror on the modern trajectories of China'], *Hérodote*, no. 116, 2005.
10 Mycenaean Greece owed its prosperity to its military superiority over its enemies — particularly the Trojans — which it acquired from such weapons.
11 The Incas' and other Andean cultures' mastery of copper and bronze was no match for the Conquistadors' mastery of iron. This contributed to the swift conquest of the Americas in the fifteenth and sixteenth centuries. See Eric Chaline, *50 Minerals that Changed the Course of History*, Firefly Books, 2012.
12 Nabeel Mancheri, Lalitha Sundaresan, and S. Chandrashekar, 'Dominating the World: China and the rare earth industry', *National Institute of Advanced Studies*, 2013.
13 'Panama Papers: from the Guatemalan Drug Queen to the "most dangerous mobster in the world"', *The Irish Times*, 9 May 2016.
14 The Watergate scandal led to the resignation of President Richard Nixon in 1974.
15 'China's First Family Comes Under Growing Scrutiny', *The New York Times*, 2 June 1995.
16 The 'four commandments' governing the policy are: 'Combine the military and civil', 'Combine peace and war', 'Give priority to military products' and 'Let the civil support the military'. Together, they form sixteen ideograms, hence the name 'sixteen characters'.
17 'China's Spies "Very Aggressive" Threat to U.S.', *The Washington Times*, 6 March 2007.
18 Interview with Hugo Meijer, currently a researcher at L'Institut de recherche stratégique de l'École militaire (IRSEM), 2016.
19 Interview with Steve Constantinides, Arnold Magnetic Technologies, 2016.

20 Ibid.
21 Scott Wheeler, 'Trading with the Enemy: how Clinton administration armed communist China', *American Investigator (Free Republic)*, 13 January 2000.
22 'Illegal Fundraiser for the Clintons Made Secret Tape because He Feared Being ASSASSINATED over What He Knew – and Used It to Reveal Democrats' Bid to Silence Him', *The Daily Mail*, 23 February 2017.
23 'Democrats Return Illegal Contribution; Politics: South Korean subsidiary's $250,000 donation violated ban on money from foreign nationals', *The Los Angeles Times*, 21 September 1996.
24 'Chinese Embassy Role in Contributions Probed', *The Washington Post*, 13 February 1997.
25 See the report 'The Asia-Pacific Maritime Security Strategy: achieving US national security objectives in a changing environment', US Department of Defense, 2015.
26 'Trends in World Military Expenditure, 2018', *Sipri*, April 2019.
27 'US Admiral Warns: only war can now stop Beijing controlling the South China Sea', op. cit.
28 'Full Transcript: acting FBI director McCabe and others testify before the Senate Intelligence Committee', *The Washington Post*, 11 May 2017.
29 'Presidential Executive Order on Assessing and Strengthening the Manufacturing and Defense Industrial Base and Supply Chain Resiliency of the United States', The White House, 21 July 2017.
30 'It's Not Buy America: admin aide on Trump's sweeping industrial base study', *Breaking Defense*, 25 July 2017.
31 'U.S. Launches National Security Probe into Aluminum Imports', *Reuters*, 27 April 2017.
32 'Interior Releases 2018's Final List of 35 Minerals Deemed Critical to U.S. National Security and the Economy', *USGS*, May 2018.
33 'Murkowski, Manchin, Colleagues Introduce Bipartisan Legislation to Strengthen America's Mineral Security', U.S. Senate Committee on Energy & Natural Resources, 3 May 2019. Refer also to the Rare Earth Element Advanced Coal Technologies Act (REEACT), championed by Lisa Murkowski, aimed at developing coal-based rare-earth extraction technology: 'Manchin, Capito & Murkowski Reintroduce Rare Earth Element Advanced Coal Technologies Act', U.S. Senate Committee on Energy & Natural Resources, 5 April 2019.
34 Interview with Ed Richardson, US Magnetic Materials Association, 2017.
35 'Xi's Visit Boosts China's Critical Rare-earth Sector', *Global Times*, 5 May 2019
36 'Commentary: U.S. risks losing rare earth supply in trade war', *Xinhua*, 29 May 2019.

37 'U.S. to Ensure Rare-Earth Supply Amid Trade War with China', *Bloomberg*, 4 June 2019.
38 'Rare Earth Elements – the vitamins of modern history', HIS Markit, 25 October 2019.
39 'The U.S. Rare Earths Saga Continues …', *Investor Intel,* 22 July 2019.
40 'Lynas and Blue Line MoU for Rare Earths Separation', *Mining Magazine,* 5 June 2019.
41 This law was broadened in 2009 to include electromagnets.
42 'Defense Science Board Task Force on High Performance Microchip Supply', Office of the Under Secretary of Defense for Acquisition, Technology, and Logistics, 2005.
43 'Exclusive: U.S. waived laws to keep F-35 on track with China-made parts', *Reuters*, 3 January 2014.

Chapter Eight: Mining goes global

1 These essential minerals included aluminium, lead, iron, copper, nickel, chrome, and zinc. Source: United States Geological Survey Data Series 140.
2 Alain Liger, Secretary General of the French Strategic Metals Committee (COMES), *'Transition énergétique: attention, métaux stratégiques!'* ['The Energy Transition: it's about strategic metals!'], The French High Council for Economy, Industry, Energy and Technology, 7 December 2015.
3 Since the production of steel and copper — two major base-metal products — was stable between 1970 and 2000, there was no real concern about the possibility of a mineral shortage. It was only in 2005 that industrial players and the press began to talk about scarcity after the sudden emergence of China on the raw materials market, which put immense strain on supplies.
4 *'Ressources minérales et énergie: rapport du groupe Sol et sous-sol de l'Alliance'* ['Mineral Resources and Energy: report from the Alliance's ground and underground group'], Alliance nationale de coordination de la recherche scientifique (ANCRE), June 2015. For further reading, also refer to Olivier Vidal's book *Mineral Resources and Energy: future stakes in energy transition*, ISTE Press Ltd, 2018.
5 Olivier Vidal, Bruno Goffé, and Nicholas Arndt, 'Metals for a Low-Carbon Society', *Nature Geoscience*, vol. 6, November 2013.
6 Ibid.
7 'The Growing Role of Minerals and Metals for a Low Carbon Future', The World Bank Group, June 2017. See also *'Métaux: les besoins colossaux de la transition énergétique'* ['Metals: the colossal needs of the energy transition'], *Les Échos*, 20 July 2017.
8 'How Many People Have Ever Lived on Earth?', Population Reference Bureau, 2011.

9 Interview with Olivier Vidal, 2017.
10 Interview with Alain Liger, 2016.
11 These fifteen metals are antimony, tin, lead, gold, zinc, strontium, silver, nickel, tungsten, bismuth, copper, boron, fluorite, manganese, and selenium. The five additional metals are rhenium, cobalt, iron ore, molybdenum, and rutile. See '*De surprenantes matières critiques*' ['Surprising Critical Materials'], *L'Usine nouvelle*, 10 July 2017.
12 'Critical Metals in the Path towards the Decarbonisation of the EU Energy Sector', Joint Research Centre of the European Commission, 2013.
13 Interview with John Petersen, 2017.
14 'Dwindling Supplies of Rare Earth Metals Hinder China's Shift from Coal', *TrendinTech*, 7 September 2016.
15 An eventuality not ruled out by Vivian Wu, who saw it as a 'fine' idea for China to keep all its rare-earth resources to itself.
16 Interview with Jack Lifton, 2016. Read also the speech by Christian Hocquard of the French Geological Survey included in a report released on 6 July 2015 on the implementation of a rare earths and strategic and critical raw materials by the Parliamentary Office for Scientific and Technological Assessment (OPECST).
17 Paul Valéry, *Regards sur le monde actuel*, Librairie Stock, Delamain et Boutelleau, 1931.
18 Donella H. Meadows, Dennis L. Meadows, Jorgen Randers, William W. Behrens III, *The Limits to Growth: a report for the Club of Rome's project on the predicament of mankind*, Universe Books, 1972.
19 Talk by Vincent Laflèche, the then Chairman of the French Geological Survey, Cercle Cyclope, 2016.
20 'China to Become Net Importer of Some Rare Earths', *Mining.com*, 2 January 2017.
21 Interview with John Petersen, 2017.
22 Ugo Bardi, *Extracted: how the quest for mineral wealth is plundering the planet*, Chelsea Green Publishing, 2014.
23 Ibid.
24 So says Christian Thomas, company founder of the metals recycling company Terra Nova.
25 Ester van der Voet, Reijo Salminen, Matthew Eckelman, Gavin Mudd, Terry Norgate, Roland Hisschier, Job Spijker, Martina Vijver, Olle Selinus, Leo Posthuma, Dick de Zwart, Dik van de Meent, Markus Reuter, Ladji Tikana, Sonia Valdivia, Patrick Wäger, Michael Zwicky Hauschild, and Arjan de Koning, 'Environmental Risks and Challenges of Anthropogenic Metals Flows and Cycles: a report of the working group on the global metal flows to the International Resource Panel', Kenya, United Nations Environment Programme (UNEP), 2013.
26 Bardi, *Extracted*, op. cit.

27 Interviews with Jean-Paul Tognet, 2016 and 2017.
28 Bardi, *Extracted*, op. cit.
29 Simon Winchester, *The Map that Changed the World: William Smith and the birth of modern geology*, HarperCollins, 2001.
30 'Hands Off Brazil's Niobium: Bolsonaro sees China as threat to utopian vision', *Reuters*, 25 October 2018.
31 'Australia's 15 Projects Aim to Break China Rare Earths Dominance', *Financial Times*, 3 September 2019
32 'Bill Gates and Fellow Billionaires Invest in AI Mapping Technology to Search for Ethical Cobalt', *Small Caps*, 6 March 2019.
33 'Japan to Import Rare Earth from India', *Reuters*, 28 August 2014.
34 'Merkel Signs Export Deal with Mongolia', *The Local*, 13 October 2011.
35 'North Korea Could Rival China on Rare Earths Reserves', *RT*, 9 January 2012.
36 The Comprehensive Economic and Trade Agreement (CETA) was signed with Canada in February 2017 and aims to facilitate investment by European businesses in the Canadian mining sector.
37 'President Trump's Interest in Buying Greenland: 5 questions, answered', *BBC*, 16 August 2019.
38 Interview with Didier Julienne, 2016.
39 '*Métaux: les besoins colossaux de la transition énergétique*' ['Metals: the colossal needs of the energy transition'], op. cit.
40 'Switch to Renewables Won't End the Geopolitics of Energy', *Bloomberg*, 21 August 2017. The author of the article is also co-authored a report with Meghan O'Sullivan, Indra Overland, and David Sandalow, 'The Geopolitics of Renewable Energy', Columbia University Center on Global Energy Policy, June 2017.
41 Interview with Vivian Wu, 2011.
42 'Luanda Names Rare Earths a Priority in Bid to Entice Beijing', *Africa Mining Intelligence*, 20 December 2016.
43 'Angola's Chinese-built Rail Link and the Scramble to Access the Region's Resources', *Africa China Reporting*, 26 February 2014.
44 The day after the embargo, a kilogram of terbium sold for €2,900. Now it is worth little more than €500. After trading at over €2,800, a kilogram of dysprosium now trades ten times lower. Source: mineralprices.com.
45 Interview with Christopher Ecclestone, 2016.
46 'California Rare Earths Miner Races to Restart Refining after China Doubles Tariffs in Trade War', *The Japan News*, 26 August 2019.
47 Ibid.
48 'Exclusive: Evidence of Chinese interests driving effort to block Stans Energy in Kyrgyzstan', *InvestorIntel*, 18 April 2013.
49 'Mountain Pass Sells for $20.5 Million', *Mining.com*, 16 June 2017.

Chapter Nine: The last of the backwaters

1 '*Montebourg veut que la France retrouve sa bonne mine*' ['France Wants to Make a Mining Comeback'], *AFP*, 22 February 2014. The statement was made during the minister's visit to the gypsum quarries of Montmorency, north of Paris in 2014.

2 '*Montebourg dote l'Etat d'un bras armé dans le secteur des mines*' ['Montebourg's Secret Weapon for France's Mining Sector'], *Les Echos*, 21 February 2014.

3 '*Macron enterre la Compagnie des mines de France chère à Montebourg*' ['Macron Shelves Montebourg's Beloved National Company of French Mines'], *Challenges*, 9 February 2016.

4 '*Emmanuel Macron préside l'installation du groupe de travail chargé de définir la "mine responsable" du xxie siècle*' ['Emmanuel Macron Presides over the Installation of a Working Group Tasked with the Definition of "Responsible Mining"'], Press Release from the French Ministry of the Economy and Finance, 1 April 2015.

5 '*Emmanuel Macron engage la démarche "mine responsible"*' ['Emmanuel Macron Launches the "Responsible Mining Initiative" in the Twenty-first Century'], *Minéral Info*, 28 March 2015.

6 '*Une réforme du code minier pour enterrer le gaz de schist*' ['Mining Code Reform to Bury Shale Gas'], *Le Monde*, 17 January 2017.

7 '*Creuser et forer, pour quoi faire? Réalités et fausses vérités du renouveau extractif en France*' ['Digging and Drilling: what for? Realities and false truths about the revival of mining in France'], *Les Amis de la Terre France*, December 2016.

8 Ibid.

9 In France, there are believed to be as many 3,500 former mines that are still polluted with heavy metals. See '*Mines: l'héritage empoisonné*' ['The Poisoned Legacy of Mining'], *France Culture*, 5 May 2017.

10 'When a River Runs Orange', *The New York Times*, 20 August 2015.

11 Ibid.

12 'Mining Report Finds 60,000 Abandoned Sites, Lack of Rehabilitation and Unreliable Data', *ABC*, 16 February 2017.

13 'Abandoned Mines and the Water Environment', *Environment Agency*, August 2008.

14 '*La ruée sur les métaux*' ['The Metals Gold Rush'], *Le Monde*, 13 September 2016.

15 Les Amis de la Terre (Friends of the Earth) encourages recycling rare metals rather than moving back to mining. As attractive as this is, and given the meteoric growth in renewable energy, I do not believe their suggestion will put a dent in the growth of the mining sector.

16 '*Emmanuel Macron préside l'installation du groupe de travail chargé de définir la "mine responsable" du xxie siècle*' ['Emmanuel Macron Presides over the Installation of a Working Group Tasked with the Definition of "Responsible Mining" in the Twenty-first Century'], op. cit.

17 'US GAO Warns It May Take 15 Years to Rebuild U.S. Rare Earths Supply Chain', *MineWeb*, 15 April 2010.
18 *'Nos déchets nucléaires sont cachés en Sibérie'* ['Our Nuclear Waste Is Hidden in Siberia'], *Libération*, 12 October 2009.
19 The two monarchs returned to Paris in July 2019 on the invitation of President Macron. See *'Contrat de convergence de Wallis et Futuna: déplacement royal à Paris'* ['Convergence Agreement of Wallis and Futuna: royal visit to Paris'], *France Info*, 6 July 2019.
20 There are in fact three kings (and three kingdoms) in Wallis and Futuna: two kings in Futuna and one king in Wallis. Following a royal dispute in 2017, only one of the two kings in Wallis is recognised in Paris.
21 The kings of Futuna are believed to be paid a much lower amount.
22 Interview with Pierre Simunek, former secretary-general of the prefecture of Wallis and Futuna, 2016.
23 Ibid.
24 French White Paper: Defence and National Security 2013, Ministère de la Défense, 2013.
25 Interview with Pierre Simunek, 2016.
26 'The Tremendous Potential of Deep-sea Mud as a Source of Rare-earth Elements', *Scientific Reports*, 10 April 2018.
27 'It's Only a Matter of Time before Deep-sea Mining Comes to Canada. We're not ready', *The Narwal*, 26 March 2019.
28 Interview with Pierre Cochonat, currently a consultant in marine geosciences, 2013.
29 'Countries with the Largest Exclusive Economic Zones', *World Atlas*, 29 June 2018.
30 'Commission on Limits of Continental Shelf Meeting at Headquarters, 11 July–26 August', Background release, United Nations (UN), 11 July 2016.
31 Max Mönch, Alexander Lahl, *Ocean's Monopoly*, produced by Werwiewas, 2015.
32 H.R.2262 — US Commercial Space Launch Competitiveness Act, 114th Congress (2015–2016).
33 Refer to websites of Space Resources Australia, Platinoid Mines Corporation and Asteroid Mining Corporation.
34 NewSpace entrepreneurs include Elon Musk of SpaceX, Jeff Bezos, the founder of Blue Origin, or Greg Wyler, the director of One Web
35 Interviews with Olivier Sanguy, editor in chief of the website Enjoy Space, and MarieAnge Sanguy, editor in chief of the journal Espace & Exploration, 2016.
36 Read the fascinating article written by the astronaut Thomas Pesquet, *'Mines dans l'espace, la nouvelle frontière'* ['Mines in Space: the new frontier'], *Les Échos*, 8 October 2017.
37 Luxembourg is hardly a novice, as it already headquarters the world's leading satellite operator SES Group.

Epilogue

1 Michio Kaku, *Physics of the Future: how science will shape human destiny and our daily lives by the year 2100*, Allen Lane, 2011. See also 'China to Build World's First Solar Power Station in Space in Next Five Years', *Newsweek*, 4 November 2019.
2 '*Photovoltaïque: les promesses des pérovskites*' ['Photovoltaics: the promises of perovskites'], *Le Monde*, 15 June 2017.
3 See Pierre-Noël Giraud and Timothée Ollivier, *Économie des matières premières* [*The Economics of Raw Materials*], op. cit.
4 See Philippe Bihouix, *L'Âge des low technology: vers une civilisation techniquement soutenable* [*The Age of Low Technology: towards a sustainable technical civilisation*], Seuil, 2014.
5 Translator's note: Originally a French term (*décroissance*), 'degrowth', refers to the downscaling of production and consumption (energy and resources), and to decoupling growth from improvement.
6 Interview with Christian Thomas, 2017.

CPSIA information can be obtained
at www.ICGtesting.com
Printed in the USA
JSHW030025201222
35166JS00002B/72

9 781950 35431